Irish Writers Against War

ALL ROYALTIES TO THE IRISH ANTI-WAR MOVEMENT

Conor Kostick is an historian based at TCD. He is the chairperson of the Irish Writers' Union and founder of Writers Against the War. His publications include the books *Revolution in Ireland* and (with Lorcan Collins) *The Easter Rising – A Guide to Dublin in 1916*. He is currently a reviewer for the *Journal of Music in Ireland*.

Katherine Moore has worked as an administrator for the Irish Writers' Centre for several years and has an extensive involvement with the community of writers, playwrights and translators of Ireland. She has worked on previous anthologies, including *Out to Lunch* and *The Whoseday Book*.

IRISH
WRITERS
AGAINST
WAR

Edited by

Conor Kostick and **Katherine Moore**

THE O'BRIEN PRESS
DUBLIN

First published 2003 by The O'Brien Press Ltd,
20 Victoria Road, Dublin 6, Ireland.
Tel: +353 1 4923333; Fax: +353 1 4922777
E-mail: books@obrien.ie
Website: www.obrien.ie

ISBN: 0-86278-825-0

British Library Cataloguing-in-Publication Data
A catalogue record for this title is available from
the British Library

1 2 3 4 5 6 7 8 9 10
03 04 05 06 07 08 09 10

Editing, typesetting, layout and design: The O'Brien Press Ltd
Printing: The Woodprintcraft Group Ltd.

Contents

Preface

I will not be enlisted into the strident ranks of anti-America. My affectionate relationship with the USA began with emigrant ancestors and for over fifty years my own experience of the United States has been animating and enriching.

So I am deeply uneasy with the vocabulary of the current America-bashers. Their language of abuse is an abuse of language – 'imperialist monsters', 'nation of terrorists', 'Christian thugs' (this from a dramatist!). To those of us who claim to have a responsibility to considered words, that abuse of words is an offence. But I will certainly not be enlisted by America and her allies into their war against Iraq; and I find it difficult to express in temperate language my anger at that prospect. Of course, emotive words are spat out by my camp too – 'blasted homes', 'starving families', 'mutilated children' – an equally dangerous vocabulary and an equally unhelpful response.

So I will say only this: that I oppose this war with a mute passion, a pain of deep anxiety that cannot find coherent articulation. And I oppose this war because I just know – every instinct insists – that there is something not-thought-through about it; something wildly disproportionate about it; something inimical to reason and reasonableness; something, indeed, that offends the notion of what it is to be fully human.

If this stance classifies me as an opponent of the US and her allies and the whole axis of terrible vengeance, then regretfully – while this distemper rages abroad – so be it.

Brian Friel (2003)

Editors' Introduction

Writing cannot stop wars, but it has an important part to play in strengthening the convictions of those who can. Indeed sometimes a poetic image can strike home more effectively than the most well reasoned political argument. This anthology has gathered contributions from Irish writers in the hope of not only providing material support for the worldwide anti-war movement, but also to make available a collection of writing that inspires and deepens opposition to war.

The anti-war movement has been accused of giving succour to repressive regimes, but while the immediate focus of this anthology is to protest against war on Iraq, there is a great deal of material here that no dictator would wish to see circulating in their country. We have included works that speak of human dignity in the face of repression, and writing that is clearly opposed to all perpetrators of injustice.

As for the idea that opponents of the war are hostile to America, we cannot express any more clearly than has Brian Friel that the opposite is true. America is a country that has made an enormous contribution to human culture. The source of its creativity, however, is not to be found in an administration that is willing to unleash destruction on one of the world's oldest cities.

We have organised the book alphabetically by author, which might seem somewhat arbitrary, but there is something fresher about this than trying to fit the writing into a formula of the sort: two poems, then fiction, then article, then back to two poems.

The response to our appeal for contributions to this anthology was overwhelming. We would like to thank everyone who sent in their works and their words of encouragement. All the contributors have waived any payment in favour of the Irish

Anti-War Movement. We regret that it was not possible to include everything that was submitted.

We would like to thank Bernadette Larkin and the other staff of the Irish Writers' Centre and the staff at O'Brien Press who gave their own time to ensuring the swift production of this anthology.

Finally, even though this is a book of protest, we also hope that it will be enjoyed for the intrinsic value of the work that it contains.

Conor Kostick
Katherine Moore

Philip Casey

Philip Casey's publications include the verse collection *The Year of the Knife – Poems 1980-1990* (1991), and three novels: *The Fabulists* (1994); *The Water Star* (1999); and *The Fisher Child* (2001), which completes *The Bann River Trilogy*. A member of Aosdána, he initiated and maintains the websites 'Irish Writers Online' and 'A Guide to Irish Culture'. He lives in Dublin. This excerpt from *The Water Star*, which was published by Picador in 1999, is reprinted by kind permission of the author and the Lisa Eveleigh Literary Agency.

from *The Water Star*

That night he woke again to the sound of weeping and snoring. He sat up and thought of Elizabeth, of her mother, and as he stared into the shadows, he felt himself float out of the ward into a much deeper darkness, where the air was chilled. As this darkness gave way to dawn he found himself looking into the cold eyes of a seagull. When he looked down all he could see was the dull grey of water, its waves rising and falling. There was something unusual on the horizon and he watched it slowly take shape. There were three tall-masted ships, their white sails full, towing a metropolis, and one by one he recognised the buildings and monuments. It was London, being towed across what he now realised was the North Sea, the London of the late 1930s, untouched by war.

On the opposite horizon, as he had half expected, another flotilla of three masted ships was towing a second city towards London. He journeyed ahead and long before he had reached it knew that it was Hamburg, prior to the war. There it was, the rollercoaster of the Hamburger Elbbrüke, with its castellated towers; the Rathaus and the Hauptbahnhof, the steam rising from the old engines as if nothing would ever change.

And yes, there they were, the early morning crowd at the Messberg market. He was full of love for them, as they bought and sold their fruits and vegetables, absorbed in the ordinary. And there again, the Jungfernsteig, with the *Alster* steamer coming into dock, people strolling along the promenade, a flag billowing above the turret of the Alster Pavilion.

The two cities sailed towards each other all day. Karl was hungry and thirsty, but he could do no other than endure. To sustain him, he let the second half of the Eighth Symphony, the final scene from *Faust*, flow through his head. It grew and grew until it spilled out of his skull and poured into the sky above the North Sea.

It was late afternoon when the German tall ships sailed up the Thames, and the English tall ships sailed up the Elbe until they locked together as one.

Jimmy woke him.

'Mr Bruckner, time to get up. Washey washies. Brekkie brekkies.'

'Eh?'

'Race you?'

'O yes,' Karl smiled. His outstretched arms ached badly, and he realised that he had been flying for most of the night.

Elizabeth came again, and he wanted to tell her about his journey above the North Sea, but he was afraid it would upset her. He wasn't mad, of that he was sure, but he had no doubt it had actually happened, not in the material, logical world, but in that realm of existence where reality was ordered and repaired.

'...*balde* ...*Ruhest du auch*,' he repeated to himself. Yes. Soon you too will rest. But there was another journey to be undertaken before he could rest. It would lead him to the spirit of his family. He thought about this on his daily walk. He realised that to be allowed his walk like this was a privilege, as many never left the ward unless in supervised groups.

He did not tell Dr Greene about his journey. The doctor had asked him about dreams, but this was a journey, not a dream. Instead they talked a lot about his childhood, at first in Erxleben and Berlin, and then in Hamburg. Karl remembered it well, but the faces of his family were blank.

And then, quite suddenly, it was Christmas. The nurses decorated the ward, and made them all wear hats, which made them look truly mad. They had special meals all day and some of the nurses smelt of brandy. The Romanian wardsmaid was tipsy and when she thought no one was looking, she kissed him, forcing her tongue into his passive mouth.

'What is your name?' he asked as she turned away, embarrassed.

'Marta,' she said without looking back.

Marta.

As it grew dark, a group of nurses sang carols.

Where had the months gone, and where was Elizabeth?

One afternoon in the New Year, he was sitting under his favourite tree when Hugh came down the steps towards him, looking apprehensive. Of course he was apprehensive. Karl remembered that he was afraid of hospitals, so to relax him, he smiled and offered his hand, which Hugh shook.

'Sit down, Hugh, it's good to see you.'

'How are you at all, Karl?'

'I'm good, Hugh. Things got a little too much for me, that's all.'

'Should we go inside?' Hugh asked, looking up. 'It's going to rain.'

'It won't rain just yet.' Karl looked up too, but into the naked branches of the tree. 'Do you know the legend about this tree?'

'A legend? I don't. What tree is it anyway?'

'It's a white poplar. Hercules won a battle, the story goes, and on the mountain there was a grove of these trees, and he made a garland of its leaves to celebrate his victory. Soon after, for what reason I forget, he descended into the place of tears and gloom, the Underworld. When he came back to this world, the top of his garland was darkened by the smoke of hell, but underneath was stained white by his sweat.'

'That's a good story.'

'If it were not autumn, I could make you a garland.'

Hugh laughed.

'Actually it's winter, Karl.'

'Is it? There are many fine trees here. Would you like a tour?'

'Are you sure it's not going to rain?'

'It will not rain. Not until we are ready.'

'Right, so.'

They walked among the trees and shrubs, Karl educating Hugh about them.

'I heard a story once,' said Hugh, 'that long ago in Ireland, the names of the trees were the same as the letters of the alphabet.'

Karl stopped and looked at Hugh in surprise.

'Hugh, you have astonished me again.'

It started to rain lightly and they walked back to the hospital.

'How is Elizabeth?' Karl asked him in the hall.

'She's grand. I suppose she told you I was frightened out of my wits about coming here?'

Karl smiled, shaking his hand.

'When are you going to marry her?'

'Ah now!' Hugh exclaimed, and he shuffled from foot to foot. Then he left.

'Will we play some music?' Dr Greene asked him at their next session.

'The *Eighth*,' Karl said emphatically.

'Is it here? Yes, it's here.'

'The final scene from *Faust*. They should be all together.'

'The ah …' said the doctor, examining the labels, 'the final scene from *Faust*. Here we are.' He looked at Karl, before putting on a batch of records. Midway through, as the choir began to sing, Karl, his eyes closed, began.

'So much destruction … the bricks piled high in the street. So high …'

'Is this Hamburg?' came a soft voice. 'Hamburg during the war?'

'This *was* … Hamburg.'

'Go on. If you can.'

'Did you know that bricks fall intact after a bomb? I've seen it in London, you know. You can take a brick from a bombed building, tap the brick with a trowel to clean it, and then put it straight into a new wall. They fall intact and clean. So many of them, lying in every possible relation to the other. It leaves whole streets like a great river of brick in flood.'

He felt himself go deeper.

'But night is falling, and soon they will be here again.'

'Who will be here?'

'The bombers, of course.'

'You are waiting for them?'

'I have been waiting for them half my life. Night has fallen. There is silence, and I am lying down, waiting for them. The bricks are uncomfortable, but I am not here for comfort.'

'You are on the street? Why not a bomb shelter?'

'There … a dog, and now another, and another, and another, howling. I can hear their howls above the sirens, and above them both I can hear the drone, getting louder … and then louder still. At last. At long last …'

The drone of the heavy bombers went on, and on.

'They are overhead for hours, and yet … I'm still waiting for the bombs to fall.'

He could bear it no longer. He had waited too long, and this time he had been sure it would happen. His tears spilled over. The record stopped, and Dr Greene put on another.

Ah.

The soft voice of the tenor, and the slow beat of the music lulled him until the harp struck and the choir soared and in his head he opened his eyes.

Everywhere was on fire. As far as he could see. On the rubble of a collapsed building three lonely men pumped water on to the mountainous, ravenous flames. Around him, thousands of people were running, their screams sucked away by the tumult of the fires and dwindling air.

A roar gathered until it deafened him and the sky darkened, the fire shining a dark red and dull white through the airborne debris until it was shut out and darkness prevailed. Then a huge fire ignited like a great spit, illuminating a young girl, her mother running after her. Karl watched, powerless, from a distance of years. The tar melted beneath her feet, and, screaming, she could not drag herself from it, her mother's fingers almost touching her; but as her mother was stuck too, they would never reach each other.

The fire consumed them.

The tar had melted beneath his feet also, and he fought to drag his shoes from it. A terrible wind came through the streets and alleyways, gathering power as it joined and multiplied within itself. It blew him almost on to his back, his shoes still stuck to the tar. But there was no air.

'I can't breathe,' Karl choked, 'I can't breathe.'

The soprano and contralto sang in harmony as he rose within the whirlwind.

Here, there was no more fire, no turmoil as with a multitude of others he rose higher and higher within the spiral until he could see the clear light of the evening star. And then he saw that the bombers had not departed, they had waited for him after all, waiting above the city they had destroyed with their incendiaries. He could feel the wind from the propellers as he moved among them, and he saw the young, tense faces of the crews as the soprano's voice rose into a hushed purity.

He wandered among the suspended planes, thousands of them, and saw that they were Lancasters and Flying Fortresses and Junkers Ju 88s and Messerschmitts, waiting for him, for others, down the years, down the centuries.

At last. At last he knew where he was. He walked among the suffocated and incinerated people of his city, blowing air between their lips to cool their boiling lungs, pouring oil over their charred bones. The young crews of the warplanes stared into infinity, intent on survival, no room in their terrified hearts to know what they had done all over Europe.

'Play "*Alles Vergängliche*",' he ordered the doctor from 1944, and he waited until the choir sang before turning to greet his family as they made their way out of the massed dead piled between the planes. The soprano's and contralto's and tenor's voices rose into the great chorus as Gertrud called out his name. Like Mama and Papa she was bloodied and in rags, covered in the torment of the storm. Jurgen was frozen in his Wermacht uniform, but he saw past all of that into the life they had shared as he embraced them.

Michael Coady

Poet, author, musician and former teacher, Michael Coady lives in his home town, Carrick-on-Suir, County Tipperary, and is a member of Aosdána. The winner of a number of literary awards and author of three books with Gallery Press, he is also known as a broadcaster and a participant in arts events at home and abroad. His collections are *Two for a Woman, Three for a Man* (1980), *Oven Lane* (1987). 'There Are Also Musicians' is reproduced by kind permission of Michael Coady and the Gallery Press.

There Are Also Musicians

Though there are torturers in the world
there are also musicians;

though, at this moment,
men are screaming in prisons,

there are jazzmen raising storms
of sensuous celebration

and orchestras releasing
glories of the spirit.

Though the image of God
is everywhere defiled,

a man in West Clare
is playing the concertina,

the Sistine Choir is levitating
under the dome of St Peter's

and a drunk man on the road
is singing for no reason.

Tim Pat Coogan

Former editor of the *Irish Press*, Tim Pat Coogan is the author of numerous books including, most recently, *1916: The Easter Rising*; *Eamon de Valera*; *Michael Collins*; *The IRA;* and *Wherever Green Is Worn*.

The State of America

To my once favourite Uncle Sam. This book is an exercise in democracy. In a democracy, the difference between a friend and a lackey is that a friend will, and is free to, sometimes tell you when they think you are going wrong. As a friend of America's for many decades, my purpose in joining the Iraq debate is to say loud and clear that while I am strongly pro-American in many fields of human activity, I am increasingly at odds with certain of its foreign policies. Over the years I have found Americans to be more open, and more generous, than most. However, American foreign policy does not always reflect that generosity of both purse and spirit. I believe, in fact, that aspects of its foreign policy are flawed and dangerous, and particularly so under the Bush regime. Iraq cannot be considered in isolation from several other factors contributing to the situation that exists at the time that I am writing.

Firstly, it is wrong to think of America as merely a big country. It is, in fact, a large continent, with a wide variety of time zones, climates, foods, cultures, and geographical and psychological features. People can, and do, lead a rich and varied life without ever leaving American shores. That is why probably only one in five American congressmen hold a passport, a statistic which when revealed to me a few days after September 11th by an experienced American diplomat, struck me as one of the most significant facts concerning the policies that led to the bombing of the Twin Towers. So many Americans have so little need to step outside their wonderful continent that foreign policy formulation is not a major concern of the people who returned those congressmen to Washington, and they, in turn, apart from exceptional circumstances like those now obtaining, are not much concerned with foreign policy issues.

To a large extent, therefore, American foreign policy is influenced by lobbies, not by your average George and Martha who, if the situation in small far-off countries obtrudes upon them at all, does so only in terms of 'Gee, I hope George Junior never has to go there.' Decoded, the unspoken, post-Vietnam, addendum is – 'and come home in a body bag.' The forces which do influence foreign policy include the major lobbies: the industrial-military complex, which was named by one of Bush's predecessors, President Eisenhower, as being a threat to America's and the world's safety; the oil interests, the Israelis, the

Cubans – and the Irish. Let it not be forgotten that there would be no Irish Peace Process, imperfect and faltering though it be, were it not for the mobilisation of the Irish in America and the sympathetic reception they received from President Clinton. Because America brought us peace, we should raise our voices now to try to prevent the inter-action of the first three lobby groups mentioned, industrial-military/oil/pro-Israel, helping to bring America to war.

Historians of the future may very well conclude that electoral procedures in Florida, where, of course, the anti-Castro Cubans are particularly influential, which gave George W. Bush the presidency, were highly dubious, to say the least of it. But what is unquestionable is the fact that the man represents big business and the American Right. The Republican Party Conference which selected him as the presidential candidate probably had more millionaires amongst the delegates than wage earners. Since coming to power, Bush has followed an anti-environmentalist, pro-oil lobby agenda, dismissing Kyoto and encouraging such anti-environmental proposals as that now currently up for decision – increased logging in the famous protected Sierra forest ranges.

What is also unquestionable is the fact that Bush has a personal animus towards Saddam Hussein: 'He tried to kill my daddy.' And post 9/11, Bush has had an unrivalled basis in American public opinion in which to take the gloves off, and start settling old scores. At the geo-political level it was inevitable that the horror of the Twin Towers would elicit a massive reaction. Washington hawks, with their eyes on a China which could well rival America in power in fifty years' time, were not going to allow it to be thought that America was a soft touch for anyone.

On a human level, the desire for retribution in the wake of what happened in New York on that September morning is understandable. I have laid a wreath at Ground Zero in the company of Irish American policemen, and spoken with the grieving Pakistani, Chinese, Irish, Bangladeshi and other relatives of those who died. Only a psychopath could remain unmoved by the evidence of man's inhumanity to man, provided by grief, the twisted rubble and the sickly-smelling, death-camp-recalling smoke curling up from the bowels of what a few days earlier had been America's twin tabernacles to commerce. Tabernacles which, literally at every level of their existence, had been inhabited by people of either Irish origin or descent.

But, as with the memories of the death-camps and the holocaust, the memory of the hallowed dead is being used for unholy purposes. September 11th is being utilised to achieve long-standing policy goals. It's not merely the Iraqi people who must suffer, it is the Palestinians, the Afghans, and God knows who else, as stock exchanges plummet and sabres are rattled from the Potomac to North Korea.

The people involved in supplying Saddam, who was known to be as brutal then as he is now, were the then vice-president George Bush, George W's father, Donald Rumsfeld, the current Secretary of Defence, and the then director of the CIA, William Casey. Bush père is on record as having urged Saddam to have cluster bombs which were a 'perfect force multiplier' against the Iranian human waves. The Americans, of course, will be using cluster bombs against Iraq should war break out.

One of the reasons, inextricably linked with the Iraqi crisis, for Arab and Muslim hatred of America, is the behaviour of Ariel Sharon's government and that of the Israeli Right. America's colossal support for its Middle Eastern ally, which is partly strategic, partly the result of the influence of the Israeli lobby, formed part of the hate-filled pressure on the controls of the

Boeing jets that flew into the Twin Towers. Ariel Sharon is responsible not merely for the massacres in Lebanon, but for the countless deaths which have flowed since his appallingly cynical Temple Mount visit, carried out on 28 September 2000, purely as an election campaign stunt, aimed at his right-wing opponent, Benjamin Netanyahu. It sparked the *intifada* that now continues, spawning, on the one hand, suicide bombers and, on the other, Israeli tanks shelling refugee camps and firing on stone throwing children. Sharon is rightly considered by many to be a war criminal, and cannot travel freely in Europe without fear of arrest.

The American public, which to a great degree depends on a media, both print and electronic, of appalling inadequacy, hardly followed events in the Middle East. No more than they were aware that in their own back-yard, Latin America, the CIA was making and unmaking undemocratic governments, some of whose practices towards their own people equalled anything that Saddam has ever done. The military School of the Americas in Panama taught what are euphemistically termed 'counter insurgency techniques', which included Nazi-style tortures and other brutal methods of military repression, to thousands of military officers from whose ranks there subsequently emerged most of Latin America's military dictators. The CIA today is actively trying to destabilise Venezualian President Chavez's regime. Again, the cause of the intervention is oil.

The cartoonist Will Dyson had a cartoon in the *Daily Herald* in 1919. It showed the statesmen of Versailles, who were setting about the creation of a new world order after another cataclysmic event, World War I. With the statesmen was a child labelled prophetically 'class of 1940'. The caption had one of the statesmen saying: 'Curious, I seem to hear a child weeping.' It is not fanciful today to imagine that one can hear the children of the future weeping.

Dragon's teeth are being sown. One of the architects of the partition of this country was Arthur Balfour, one of those Versailles statesmen. It was he who made the Balfour Declaration, giving the Jews a homeland, and it was he who helped to create the country we know as Iraq by taking eight hundred miles of desert and baptising it. What he didn't know was how much oil wealth that desert would contain. Because of it, America, whose foreign policy in the region has been to step into Britain's imperial shoes, has overthrown governments, most notably that of Prime Minister Muhammed Mossaddeq in Iran. Mossaddeq had toppled the Shah, but the US re-installed him, thereby leading to the instability that brought the fundamentalist Ayatollah Khomeini to power. It was the Reagan-Bush Snr. team which manipulated the US hostage situation so that the Iranians only released the hostages *after* the anti-pollution environmentalist Jimmy Carter was defeated, to the delight of the oil lobbies. They subsequently caused America to back Saddam Hussein in Iraq's war with Iran, supplying him with weapons of mass destruction in the process. These included chemical, biological and nuclear weaponry. Saddam was supplied with anthrax and other toxins, and bacteria including botulins, e. coli and, ironically, an Israeli-developed strain of West Nile virus. The Americans made possible the Al Atheer nuclear complex which the Israelis bombed in 1992.

We no longer blink an eye at 'covert operations'. The Americans blithely cross national boundaries these days to assassinate enemies who have not been brought before judge or jury. The treatment of the Al Qaeda prisoners captured in Afghanistan is as bad as anything done by the British in Northern Ireland after internment was introduced in 1971. The Dublin government was forced by public opinion to take London to the Court of Human Rights in Europe over these techniques, but what forum exists to test the activities of the leader of the free world? In fact, the Americans are refusing to sign up for such a forum.

I do not propose to attempt to examine any pro-Saddam argument. As far as I am concerned, he is outside the Pale of decent human behaviour. But the women and children of Iraq are not and Saddam should not be used as a pretext for departing from the standards of decent human behaviour, which we would normally look to America to uphold.

John F. Deane

John F. Deane was born on Achill Island in 1943; he founded *Poetry Ireland* and *The Poetry Ireland Review*, 1979. His poetry includes *Christ, with Urban Fox*, a collection translated into several languages and *Toccata and Fugue, New & Selected Poems* (2000). Blackstaff Press published his novels *In the Name of the Wolf* (1999) and *Undertow* (2002), as well as a collection of short stories *The Coffin Master* (2000).

Suffer the Children

Noon. Spring-time. Again that hushed and cherry-blossom
first-communion purity, the smaller birds
insistent in their mating-songs, and the word *grace*
hovering like a blue-day sea-mist everywhere.

In my dream the children, loose-limbed and
    clamouring, irritate
the Christ; but today they are staring out at us
from make-shift beds in make-shift wards, the children

who have been washed over and over in the Lamb's blood.
South of Jerusalem Israeli tanks
have been rumbling into Beit Jalla;
a child's pink ballet-shoe
lies in the rubbled buildings of Jenin;

I remember, on the school-room wall, a painting,
Christ's hand gently on the head of a child,
John, the apostle, bustling others away, and underneath,
the words *Suffer the little children.*

Upstairs, back of the cupboard, the Start-rite box,
a pair of shoes, tiny pink hearts
on the half-scuffed toes; kept,
treasured, and forgotten. But remembered

a night, an almost tropical
underwater funk and desperation, kettles, steam,
the child (named Mary, cherished)
exhausted in the labour of breathing; till,

panicked at last into where we should have been,
we raced through the city coloured death,
streetlight like tainted moonwater, the actual world
uninhabitable; a nurse

snatching the motionless body, rushing her
out of our hold, leaving
one blue-wool shoe, discarded, to be stuff
of nightmares. A tap

dripped loudly
in a porcelain sink; we waited,
understanding
whirling round and round in the sink-hole.

A prayer may be (how can I say this?) this
emptiness, this waiting, incomprehension, this
imponderable hanging – breathless –
on the mercy of God.

Out on the lip of the cliffs, with the vast Atlantic
breathing out and breathing in on the beaches
    and shingles below,
the small hard grass-humps of the lost
babies, those

who had chosen to die in the flooding waters
of birth, those
we could not cope with, nor understand,
nor pray for, nor pray to –

cast at once out of our ken to the world's edge.
Raise
small crosses now, small
patterns of white shore-stones, and beg

forgiveness, for our ignorance, our innocence, the utter
foolishness of our wars,
there where the fulmar cry, where flotsam ends,
where the hurt seal comes flopping out of ocean

to find rest.

Raymond Deane

Brought up on Achill Island, Co. Mayo, Raymond Deane now divides his time between Dublin, Paris, and Ceret in French Catalonia. A classical composer, member of Aosdána since 1987, he has also published the novel *Death of a Medium* (1992), and many essays and articles on cultural and political issues. In 2002 he was a founding member of the Ireland Palestine Solidarity Campaign.

War and Despair

after Andreas Gryphius (1616-64)

Now once again comes utter woe:
The mercenary mob, the trumpet brash,
The blood-caked sword, the cannon's crash
Have laid all thrift and effort low.

Thrones lie in dust, the strong are hacked apart;
Towers collapse and churches blaze,
Virgins are ravished; where we gaze
Are death and plague, piercing the heart.

The city streets are cleansed with blood,
The river's current changed to mud
As piled-up corpses block its flow.

And yet I cannot speak of what is worse
Than death and plague and famine's curse:

The human spirit robbed of hope.

Theo Dorgan

Born in Cork, 1953, Theo Dorgan has published a number of pamphlets and three books of poetry: *The Ordinary House of Love* (1991), *Rosa Mundi* (1995) and *Sappho's Daughter* (1998). He is editor of *Irish Poetry Since Kavanagh* (1995) and co-editor of *Revising the Rising* (1991), *The Great Book of Ireland* (1991), *Watching the River Flow* (2000) and *The Great Book of Gaelic* (2002). Editor, documentary scriptwriter, broadcaster on radio and television, he is a member of Aosdána. 'Of Certain Architects, Technicians and Butchers' is from the collection *Rosa Mundi*, published by Salmon Press in1991.

Of Certain Architects, Technicians and Butchers

I am the belly of great armies
I battle the ages in my fear
I am the horror in the newborn child
And the horror in its mother.
Who but I built Ulm cathedral?

I am the great cathedrals of pure thought
I am the frozen wave of the Carpathians
I am the sword that cleaves the knot forever
And the knot itself that closes around the sword.
Who but myself makes Alexanders weep?

I am the famine when Alexanders weep
I am the sand that built and swallowed Carthage
I am the Hydra drinking Stalin's blood
And the blood that doused the flames of Dresden.
Who but I could darken the air with engines?

I am the engines and the fire at heart
I am the garden and gardener at Treblinka
I am the wolf who howls in Katyn forest
And the forest itself howling in the wolf.
Who but I would storm the moon?

I am a forest of great armies
I am the knot that twists the child
I am the sword turns in engines
I am the wolf in the cathedral.

I pace behind mountains, turning the days over,
I wait for the dark star that will shine when I cross.

Kevin Doyle

Kevin Doyle is a short-story writer from Cork. His stories have appeared in a wide variety of publications including *The Cúirt Journal*, *Pulse Fiction* (1998), *Snapshots (1999)* and *The Stinging Fly*. In 1997 he won an Ian St. James Award with 'Do You Like Oranges?'. Currently, he is completing a novel due to be published in 2004. 'Tear Duct Capacity' previously appeared in *The Burning Bush 5* (Galway, 2001).

Tear Duct Capacity

To begin with I should tell you that this story unfolded before my eyes in a remote area of the world, in the eastern part of a small island that was the subject of some un-neighbourly activity at the time. Indeed I was conscious during the entire week that I spent on this island that the world was still a large and undiscovered place. In my line of work one grows accustomed to how small the world is becoming – with international travel getting to be what it is, with the seemingly unquenchable demand for news, with satellite TV and radio, with pictures flashing over the wires in minutes; all these things seem to have brought us closer. Yet on that island for one of the first times in over a decade I felt I had left the world behind. I was in a news warp, for if one wanted news (and there are plenty of young pen-slingers out there now trying to make their name) here was the place to be. Some amazing things were in train, incredible pictures were to be had and events to be reported on, if one was so inclined. Yet no one was. I was alone.

My exact reason for being there underlines this. Suffice to say that at the time I was at a high point in my journalistic career. I had done a series of features on the Iran-Iraq war that had received a great deal of praise. As a reward I was given an 'at-large' assignment attached to the Bureau's office in Bangkok. About two weeks into this, I decided to do a feature on rain-forest conservation. I felt it was a reasonable idea since a story on the rainforest will always find an audience. My plan was to go to Irian Jaya, but my first port of call was this island. A project, in part funded by the environmental organisation, Wild International, was underway there. I made my travel inquires, arranged tickets and set off. Twenty-four hours later I was in a camp on the side of a small mountain at the eastern end of the island.

I spent a number of days with the project and was made very welcome. I enjoyed the ease of the assignment and in particular the coolness of the hills, eventually making the acquaintance of the local authorities who regarded me from the beginning as a champion of their cause – though why exactly I never found out. One day as they were about to set off in convoy – they never travelled any other way – the local police chief called me aside and said, 'I have something that will interest you, Bob. It's down on the coast. Come with us – you'll be able to go for a swim too.'

The actual event I'm now going to tell you about took place in a former hotel in a former holiday resort on the island. It had long since been abandoned – the hotel, I mean. The mosquito screens were gone, the windows were broken and the pool was partially filled with refuse and earth. Only at the far end of the grounds, away from the main building, was there a reminder of the grandeur that once must have been – a long marble terrace, colonial in style. From there one could see the entire bay, its blue expanse broken only by patches of shallower, turquoise water.

After we parked our vehicles and met the Head Administrator – as he was introduced – we were led to a small bare room at the rear of the hotel. It was here that I first saw Suwondo P. He was suspended by his hands from a cross beam in such a way that his arms had been forced to the rear of his body plane and then up – a contortion that thrust him agonisingly forward into our immediate view. I say 'our' view because at no time during the next five days was I ever left alone with him – there was always a guard at hand.

It was unbearably hot inside the room, which had an unwashed dilapidated appearance. There was dust on the floor, some chairs in the corner and a small improvised table near the only window. Suspended among this, Suwondo P looked bereft of hope and energy. He appeared completely lifeless until I went close. Then he opened an eye and attempted to speak – that is, his mouth opened. But he couldn't or didn't make any sound. Oddly enough what I recall even now is the abiding smell from sweat and urine which I realised had accumulated on his body over a number of days.

I was given only the barest details about Suwondo P's background – that he was a student of some form and that he had been captured late at night leaving a house. Apart from this information it was never explained to me what exactly it was he did wrong and I must admit now that I never bothered to ask. Suffice to say that events in this remote area of the world have little to do with right or wrong.

Soon after our arrival a guard approached Suwondo P and took hold of his shrunken penis. He applied a small compression-type tubing clamp – tiny vice-grips that can be hand-tightened to give an effective seal. After this Suwondo P was slapped awake and water was placed in front of him – a liquid he drank with sluggish effort. Gradually the captive revived. His eyes, which he seemed barely able to open at first, stayed fully open. To me he seemed dazed and unsure of his predicament as if he was expecting to be dead but wasn't. Only after he had regained full consciousness did it occur to him what might be happening. He took the water they gave him but at one point, mid-gulp, he tried to look down – perhaps he had attempted to urinate, I don't know. In any case, a look of incomprehension crossed his face. He struggled for a moment, attempting to break free, but he was too weak.

After this the room became still. A further few minutes passed during which time Suwondo P's body relaxed – that is, he stopped fighting the web of ropes that held his body in check. He hung there, his face turned downwards as if he was studying the floor. Then the police chief nudged me. 'Look,' he said, but I couldn't see anything. I heard a snorting sound as if Suwondo P was choking – only then did I realise he was crying.

Perhaps this seems like a predictable outcome. But I assure you this was no ordinary crying. As his captors soon realised and pointed out, Suwondo P was crying to save his life. Having drunk nearly two litres of water – ' He can have what he wants,' they said – Suwondo P was now fully revived. In some way, he either saw or determined what they were doing to him, what death they had in mind. Water would accumulate in his bladder, which would eventually rupture, causing painful death.

Suwondo P began a sort of crying that I had never ever seen – and I had seen a lot by then. I was in Iran in '78 when the earthquake there killed over 25,000. Those people can cry, such crying, I'd never heard anything like it – wailing all day and night. But this was different. This was a steady, concentrated outpouring of water from the eyes that soon turned the floor around Suwondo P's feet into a wet puddle. His captors, so bemused at first, so confident in their excess, watched in horror. No matter how much water they poured into Suwondo P, he managed to get rid of as great a quantity. By sunset of that first day the Head Administrator was seething with rage, all the more so because of my presence.

A couple of years later I broached the matter with a friend of mine, a physiology researcher at The John Hopkins University Hospital near Washington DC. She did some computer modelling for me that examined tear-duct capacity, bladder size and liquid flow in the body. From this work she concluded that what Suwondo P did was beyond conceivability. But believe me it did happen. For four nights and five days water was drawn and forced into Suwondo P. Immediately after, he cried copiously, relieving himself of the need to urinate. How many litres in all I never actually figured – at least four and maybe even six per day. For a period on the third day, half-delirious from lack of sleep, I even wondered if his captors were going to lose – though in actual fact this was an impossible outcome. After all, with his penis clamped there would be a natural build up of toxins. In the end that would kill him.

During those next days and nights, resigning myself to seeing the event out, with a mixture of admiration and disgust, I watched Suwondo P struggle. In one way I was unable to comprehend how he could put on the show he did – against such overwhelming odds, why would he bother? His captors had only other, more foul ends planned if he ever won out against them – which he didn't, thankfully. When he did die, near sunrise on the fifth morning, the sky was a beautiful red. There was no jubilation among his captors because by then the event had become quite revolting to us all: Suwondo P hung there, the trunk of his body distended like a bloated ugly ball.

So there you have it – Suwondo P's story. When I did eventually get home – I was living in Washington at the time – I made an effort to have a feature done on him. I even gathered some information on other cases from other parts of the world to impose some balance on my perspective. But it didn't matter. The case of Suwondo P never saw it to press. Around the same time Ronald Reagan was shot and injured (by a madman as it turned out) and that took precedence over everything else.

Roddy Doyle

Roddy Doyle lives and works in Dublin. He is the author of six novels, the most recent of which is *A Star Called Henry* (1999), and several plays, screenplays and books for children. His most recent published work is *Rory and Ita*, a memoir of his parents. Extracts from *A Star Called Henry* (Jonathan Cape) are used by permission of The Random House Group Ltd. (London U.K.).

The following two extracts are from my novel, A Star Called Henry, *published in 1999. The narrator, Henry Smart, has been in the GPO, in Dublin, during Easter Week, 1916. In the first extract, he comes out of the building and, after five days of fighting, sees the enemy for the first time. The second extract is set five years later, and Henry sees his name on a piece of paper – his death sentence. Neither extract explains anything, but both came to mind as I began to think about the prospect of war in Iraq and my feelings about such a prospect.*

from *A Star Called Henry*

The gate was open and we were out. Across Henry Street. In single file. A man fell in front of me. I jumped. His body was riddled with bullets meant for my legs. I landed awkwardly but terror gave me back my stride and now Paddy was in front of me. The men ahead of us threw aside sections of our barricade and we ran through the gap onto Moore Street. And at last I saw the enemy. At the top of the street. At last, the khaki uniforms. We broke into two lines, to the left and the right, and kept running at them and it suddenly seemed very quiet and then the noise was deafening and trapped by the walls of the street, and I heard nothing extra but Paddy fell in front of me and he was dead and his brain and hair were on my jacket and hands and I kept running, and The O'Rahilly had been hit but he kept going, he ran in a zigzag that brought the bullets to him, and Felix fell and I left him behind me, and I was the first, the only man left upright on the street and I could see the bullets, the air was packed with them and, for a fragment of a second, I could think and I jumped at a doorway and hid.

Four or five men were crawling and diving, trying to get out of the hail of bullets. The rest were on the ground; their blood was already like an oilcloth on the street. Paddy's scalp was on my hand, as if I'd beaten his head with the wooden leg. I couldn't cry. I couldn't see him properly back there but I could see that Felix was dead as well; the bullets were still going into him but he didn't know it.

The bullets chipped at the doorway, eating away my hiding place. I abandoned my rifle, put the leg in its holster and slithered out. I hugged the house wall and crept back towards Moore Lane. Every bullet ever made flew up that street, at me, at my feet, at my head. They gouged paths in the wall an inch above me. They fractured, made powder of the pavement right beside me. But I kept going, inch on inch, my face to the wall, inch inch inch, and I could see the corner and I was around it. And up. I ran at a door and it broke in front of me. I fell into the hall and heard screams from upstairs. I got away from the door, in to the parlour window in time to see the second wave of men running onto the lane. And there was Plunkett, held up by two men, trying to hold his sword up, one spur hanging crookedly and alone from one of his boots. He saw me. He stopped and made the two men stop in the sea of bullets, and shouted.

–Come out and fight, you cowardly cur!

And he was gone. Connolly was brought past on a blanket, carried by four kids. I ran to the door, back out to the street. I could see men in the alleys and doorways but there were many more men on the ground. I ran to Cogan's, the grocer's on the corner. And through it – the smell of boiling ham – with the other survivors. To a cottage in the yard behind. And there was crying, wailing and a dead girl in the hall, face up on the earth floor, shot in the head – by one of us, there were no other bullets here. Men lay on the ground. I could hear nothing. But I could think. I was catching up. I sat against the cottage wall.

Paddy was dead. Felix was dead. I waited to feel something.

* * *

He slid a piece of paper across the desk. I turned it and read.

–Know him?

–Yes.

–Can you handle it yourself?

–No, I said. –I don't think so.

–Fine, he said. –I'll get someone else. But I'll leave it twenty-four hours. How's that for fair play?

I looked at the name again. Smart, Henry. I slid it back across the desk to him.

–It's not your writing, I said.

–No, he said. –You wouldn't be on my list at all. Even though you've let me down.

I nodded at the paper.

–Why?

–Well, he said. –If you're not with us you're against us. That's the thinking. And there are those who reckon that you're always going to be against us. And they're probably right. You've no stake in the country, man. Never had, never will. We needed trouble-makers and very soon now we'll have to be rid of them. And that, Henry, is all you are and ever were. A trouble-maker. The best in the business, mind. But –

He opened a drawer and knocked the slip of paper into its mouth.

–It'll stay there for a while yet. Now, get out. Before I'm seen being nice to you.

I stood up.

–Is that the only piece of paper? I asked.

–I wouldn't know.

–I'm dead.

–Yes.

–Because I'm a nuisance.

–Because you're a spy.

–Oh I said. –Fine. Were any of them really spies, Jack?

–You killed plenty of them yourself, he said. –Of course they were.

Dave Duggan

Dave Duggan lives in Derry, where he writes dramas for radio, stage and screen. He wrote the screenplay for the Oscar-nominated *Dance Lexie Dance* and he is currently writing a stageplay, *Spike and the Asteroids*. *The Shopper and the Boy* is a Sole Purpose Production.

from *The Shopper and the Boy*

There were no lullabies in the trenches. No sleeping, with the scurrying of the rats, the dankness of the water, the smell of death and gas everywhere. How could you sleep with that? When we marched in answer to our country's call, did we know our sacrifice would be a bloodbath?

I am the last Old Contemptible.
For the first time ever I was alone at the Cenotaph and the
Last Post was played for me.
The drizzle and the drone of the trumpets hemmed me in.
I looked at the wreath and stepped back, putting one foot
Firmly in the grave.
I'll be dead by Christmas.

I have a son in Canada.
His children sound like television to me.
My daughter in England, she'll come home to bury me and cry.
And return to England, leaving no one to tend my grave.

But I will make no surrender to the bleak bliss of death.
I will rage quietly to the end though that end will be a blessed
relief.
A relief from the shells and the groans.
And my friends calling to me.

I hear them.
Tommy Watson, Vernon Lumley, Cecil Harding.
I am coming soon.

Tommy Watson – he always had to be smoking something.
Always had a smoke of something.
And he shared it.
We shared everything.
Food. Shelter. Fear.
The stench of fear makes me tremble even now.

Tinnitus, the doctor said.
Twerp.
The noises in my ears were old age, he said.
The fool didn't know that they had the freshness of youth,
The rawness of boys out on a spree
and the bitterness of young men dying
Like flies swatted lazily in the summer.

The noises got worse after Tillie died.
Then Thiepval roared in my head
And the slime of the Somme slurped around.
But I marched on,
Each year at the Cenotaph as my last comrades faded
around me
Until now I am finally alone.

I march on even now, always marching
Ceaselessly slogging forwards and downwards.
Down to death and waste.

I came to manhood in waste.
Tommy Watson scythed by shrapnel, Vernon Lumley gassed,
Cecil Harding on a barbed wire spit.
My God left me that day and never returned.
The horrors drove him out and the shame kept him away.

Will they be there? Will they have marched on yet further?

No! They'll be there.
Tommy, Vernon and Cecil,
And Tommy will have a smoke of something, like always.

I will make no final surrender.
I will make a reunion and I will go forward to my comrades,
Ever forward to my comrades.
And Tillie.
And the quiet! Oh! the sweet quiet!

Terry Eagleton

Terry Eagleton lives in Dublin. His numerous works cover philosophy, literary theory, biography and drama. He is Professor of Cultural Theory and John Rylands Fellow at the University of Manchester. *What Is Fundamentalism?* appeared in *The Guardian* on 22 February 2003 and appears by permission of the author and Guardian Newspapers Ltd.

What Is Fundamentalism?

There are two things desirable for fighting fundamentalists. The first is not to be one yourself. The US government's war is somewhat compromised by the fact that it is run by scripture-spouting fanatics for whom the sanctity of human life ends at the moment of birth. This is rather like using the British National Party to run ex-Nazis to earth, or hiring Henry Kissinger to investigate mass murder, as George Bush has recently done by appointing him to inquire into the background to September 11th. Fundamentalists of the Texan stripe are not the best placed to hunt down the Taliban variety.

The second desirable thing is to know what fundamentalism is. The answer to this is less obvious than it might seem. Fundamentalism doesn't just mean people with fundamental beliefs, since that covers everyone. Being a person means being constituted by certain basic convictions, even if they are largely unconscious. What you are, in the end, is what you cannot walk away from. These convictions do not need to be burning or eye-catching or even true; they just have to go all the way down, like believing that Caracas is in Venezuela or that torturing babies is wrong. They are the kind of beliefs which choose us more than we choose them. Sceptics who doubt you can know anything for sure have at least one fundamental conviction. 'Fundamental' doesn't necessarily mean 'worth dying for'. You may be passionately convinced that the quality of life in San Francisco is superior to that in Strabane, but reluctant to go to the gallows for it.

Fundamentalists are not always the type who wave one fist in the air while thumping the table with the other. There are plenty of soft-spoken, self-effacing examples of the species. It isn't a question of style. Nor is the opposite of fundamentalism lukewarmness, or the tiresome liberal prejudice that the truth always lies somewhere in the middle. Tolerance and partisanship are not incompatible. Anti-fundamentalists are not people without passionate beliefs; they are people who number among their passionate beliefs the conviction that you have as much right to your opinion as they have. And for this, some of them are certainly prepared to die. The historian AJP Taylor was once asked at an interview for an Oxford Fellowship whether it was true that he held extreme political beliefs, to which he replied that it was, but that he held them moderately. He may have been hinting that he was a secret sceptic, but he probably just meant that he did not agree with forcing his beliefs on others.

The word 'fundamentalism' was first used in the early years of the last century by anti-liberal US Christians, who singled out seven supposed fundamentals of their faith. The word, then, is not one of those derogatory terms which only other people use about you, like 'Quaker' or 'Fatso'. It began life as a proud self-description. The first of the seven fundamentals was a belief in the literal truth of the bible; and this is probably the best definition of fundamentalism there is. It is basically a textual affair. Fundamentalists are those who believe that our linguistic currency is trustworthy only if it is backed by the gold standard of the Word of Words. They see God as copperfastening human meaning. Fundamentalism means sticking strictly to the script, which in turn means being deeply fearful of the improvised, ambiguous or indeterminate.

Fundamentalists, however, fail to realise that the phrase 'sacred text' is self-contradictory. Since writing is meaning which can be handled by anybody, any time, it is always profane and promiscuous. Meaning which has been written down is bound to be unhygienic. Words which could only ever mean one thing would not be words. Fundamentalism is the paranoid condition of those who do not see that roughness is not a defect of human existence, but what makes it work. For them, it is as though we have to measure Everest down to the last millimetre if we are not to be completely stumped about how high it is. It is not surprising that fundamentalism abhors sexuality and the body, since in one sense all flesh is rough, and all sex is rough trade.

The New Testament author known as Luke is presumably aware that Jesus was actually born in Galilee. But he needs to have him born in Judea, since the Messiah is to spring from the Judea-based house of David. A Messiah born in bumpkinish Galilee would be like one born in Gary, Indiana. So Luke coolly invents a Roman census, for which there is no independent evidence, which requires everyone to return to their place of birth in order to be registered. Since Jesus's father Joseph comes from Bethlehem in Judea, he and his wife Mary obediently trudge off to the town, where Jesus is conveniently born.

It would be hard to think up a more ludicrous way of registering the population of the entire Roman empire than having them all return to their birthplaces. Why not just register them on the spot? The result of such a mapcap scheme would have been total chaos. The traffic jams would have made Ken Livingstone's job look positively cushy … And we would almost certainly have heard about this international gridlocking from rather more disinterested witnesses than Luke. Yet fundamentalists must take Luke at his word.

Fundamentalists are really necrophiliacs, in love with a dead letter. The letter of the sacred text must be rigidly embalmed, if it is to imbue life with the certitude and finality of death. Matthew's gospel, in a moment of carelessness, presents Jesus as riding into Jerusalem on both a colt and an ass – in which case, for the fundamentalist, the Son of God must indeed have had one leg thrown over each.

The fundamentalist is a more diseased version of the 'argument from the floodgates' type of conservative. Once you allow one motorist to throw up out of the car window without imposing a lengthy prison sentence, then before you know where you are every motorist will be throwing up out of the window all the time, and the roads will become impassable. It is this kind of pathological anxiety, pressed to an extreme, which drove the religious police in Mecca early last year to send fleeing schoolgirls back into their burning school because they were not wearing their robes and headdresses, and which inspires family-loving US pro-lifers eager to incinerate Iraq to gun down doctors who terminate pregnancies. To read the world literally is a kind of insanity.

Aubrey Flegg

Aubrey Flegg has written two novels for young readers: *Katie's War*, which won the Peter Pan award presented by IBBY Sweden, deals with the Irish civil war, and *The Cinnamon Tree*, about landmines, both published by The O'Brien Press. The first book in his forthcoming *Louise* trilogy will be published soon.

Landmines and Child Soldiers

from *The Cinnamon Tree*

Thirteen-year-old Yola Abonda is grinding corn when she hears her young cousin, Gabbin, calling for help. Managu, their magnificent bull, leader of the herd of cattle that he was supposed to have been watching, has wandered up onto the hill. In his agitation Gabbin calls to Yola, 'The demons will get him.' But this is modern Africa and the demons he refers to are landmines, made by arms manufacturers in the 'developed' countries. In this extract Yola sets out up a forbidden path to try to locate the bull for Gabbin.

Yola forced herself to breathe slowly and listened for the great bell that Managu wore around his neck. Yes, there! Surely that was it, down there below the path.

She ran along, stretching to see over the bushes. The bell clanged somewhere below her now. She needed to be able to see; if only there were a rock or something she could climb on. All at once the path widened and there was a tree spreading a pool of shade. It was a cinnamon tree, planted long ago by the white people for its spicy bark, which they used to flavour their food. People had rested here, a cigarette packet and a couple of Coke cans made that clear. At any rate the tree was what she had been looking for, she could climb it and see how to entice the bull to safety. The tree stood only a step from the resting place. She measured the distance to the lowest branch with her eye – it would be an easy jump and she was as agile as a monkey in a tree. She stepped off the path, treading lightly, and coiled herself for the jump. She didn't remember it afterwards, but as she jumped, the ground gave under her ever so slightly; there was a tiny click.

For an eternity, in the flame of forty sunsets, she rose, thrown up and rag-dolled against the branch above. The upward blast from the landmine she had stepped on suspended her for a second, shook her as a terrier shakes a rat, then dropped her, pierced by her own bones, into the smoking pit where the mine had been laid. Spirits crowded her now, spirits of the dead jostling with spirits of the living, all fighting over the young life that she could feel pumping out of her before she lost consciousness. Her only other sensation, and the only one she would remember afterwards, was an overpowering scent of cinnamon.

It is Gabbin in the end who saves Yola's life, though she loses her leg.

* * *

In terms of inhumanity, the use of children as soldiers is on a par with the indiscriminate use of landmines. It is common practice to make children, as young as eight or ten, commit some atrocity, often against their own family. Then, unable to return home, and plagued by shame, they are further de-humanised with drugs and with myths of invincibility. This extract comes towards the end of The Cinnamon Tree. *Gabbin has been taken as a child soldier and has been forced – as he believes – to shoot his godfather, his beloved Uncle Banda. It is a moonlit night. Yola and Banda have penetrated the minefield that protects the boy-soldiers' camp by using a trained sniffer dog. She just has started down the slope of a gully cut into loose gravel when she sees Gabbin above her.*

[Yola] could see [Gabbin]. He had moved out from the shelter of the bushes and stood recklessly exposed against the moon-glow of the sky. She recognised the voice but it was twisted out of shape, somewhere between triumph and despair.

'I am Yola,' she called. 'Why do you stand in the open like a fool, Gabbin Abonda, where you can be shot!'

'You cannot shoot me. You are the fool. Your bullets will bounce off me. I have strong medicine.'

'What has given you this power, this medicine, mighty boy?'

'I have killed. I have this medicine because I have spilled the blood of my own kin.'

'Do you not recognise my voice, Gabbin Abonda?'

'Go away, I do not know you. I do not know myself. I have sold my soul by what I have done, all I can do is go on.' The boy's voice was rising. 'I can kill you ... nothing matters.'

Yola realised she was losing him. She'd never looked down the barrel of a rifle before. She must enter his world or perish.

'Mighty boy, there are those who still have power over you.'

'There are none!' he crowed.

What nonsense had he been fed? His voice was certain now, arrogant even, but she persisted.

'Those whose lives you have saved have power over you.'

'You are a girl, you have no power over me. I have saved no other life.'

Yola tingled – was there the tiniest hint of uncertainty when he said '... no other life?' She thrust for home.

'Yes, there is. There is another one here. It is your godfather, Banda.'

'He can't be, he is dead! I killed him!' Gabbin's voice was rising to a scream. Then, above her, Yola heard her uncle's voice boom out.

'Gabbin, I am your godfather. My name is Banda. I am the other whose life you saved. You aimed to miss me, you know that. Because of you, I live.'

Silence. Yola prayed that no one would move. Gabbin would need time to absorb this. But suddenly there was movement above her. Pebbles bounced and cascaded past. Banda was, even now, coming down the slope. She looked up: there he was, in stark relief against the white of the washout. This was folly. Yola knew how Gabbin's mind worked and she saw the movement as he swung his weapon to cover the new target.

'You are a ghost, Banda.' The voice was hysterical now. 'My bullets will pass through you. Look!'

'Look at the gravel, Gabbin,' Yola screamed, but her voice was shattered by a burst of gunfire. She could see the spurts of the striking bullets as they swept towards the figure above her. The muzzle flashes burned the corners of her eyes. She was next in the line of fire, but something heavy landed on top of her and swept and slid her down to the gully bottom; it was Fintan [her Irish friend]. The firing stopped. She looked up to where her uncle lay, a black crucifixion on the gravel slope. Irrational fury filled her; the boy had not let her finish her sentence.

'Look at the gravel, Gabbin, no ghost would leave footprints like these!' she shouted.

The gouged prints stood out, shadowed in the moonlight. Sand trickled from them.

Celia de Fréine

Celia de Fréine is a poet, dramatist and screenwriter. Her most recent play was *Nára Turas é in Aistear* (2000). She has also published a collection of poems *Faoi Chabáisti is Ríonacha* (2001). 'Eisiúint Rialtais' was first published in *Faoi Chabáistí is Ríonacha* by Cló Iar-Chonnachta; it appears by kind permission of the author and publishers.

Eisiúint Rialtais

Clampar i gcónaí
sa champa.
Troideanna teorann
i dtaobh buncanna.
Beartanna bia nach roinntear
go cothrom riamh.

Deirtear gur mar an gcéanna iad
na héadaí uilig, ach aithníonn tú
an stróiceadh a dhearnáil tú,
an tsáil a chas tú,
lúbóg a mhaíonn trí ghreim déag
ar an dá fhoirceann –

d'uimhir ámharach,
rud eile atá tábhachtach,
mar aon le haghaidh chróga
a chur ort féin – gan ligean
do na leanaí a fheiceáil
go bhfuil tú buartha.

Tig leat do mhodhanna féin

a mhúineadh dóibh:
braillín a dheisiú –
le huaim fhrancach,
uaim reatha is fáithime,
nó bealach ar bith is mian leo.

Mark Granier

Mark Granier was born in 1957. He has an MA in Poetry/Creative Writing from Lancaster University and currently works as an Education Officer in The James Joyce Centre in Dublin. His poems have appeared in numerous journals and magazines. 'When' is from his first collection *Airborne*, published by Salmon in 2001.

When

When the sky comes down to earth too soon
and we're woven into light's immaculate shroud;
when the black light seeding all our bad dreams blooms
in a spine of smoke bearing aloft the brain-cloud;

when one dies and in that breath millions more
are furnace-fanned to ashes that will blow
wherever the winds rage, burned-out spores
settling, out of the fuming skies, like snow;

when all those seeking Heaven's draughty halls
find the conflagration had to spread,
that angels with singed wings have fled their stalls,
leaving behind the dead to count the dead,

clouds will roll back, a full moon mirror waste
and time do what it always does: erase.

Robert Greacen

Robert Greacen was born in Derry in 1920. His poetry collections include: *The Undying Day* (1948), *A Garland for Captain Fox* (1975), *Young Mr Gibbon* (1979) and *Carnival at the River* (1990). His *Collected Poems 1944-1994* won the Irish Times Literature Prize in 1995. 'Soldier Asleep' first appeared in *The Undying Day* published by Falcon Press, 1948.

Soldier Asleep
after Rimbaud's 'Le Dormeur du Val'

Down in that green valley a river tumbles,
Rollicking sharply past the ragged water plants.
Stealthy sun creeps over the proud-backed hill,
Filling the valley with an iridescent foam of light.

A young soldier sleeps – with head flung back,
Lips half-parted – among the purple-flowered mint.
He lies loosely on the short, speared grass,
Unprotected from the dominant glare of noon.

Sleeping, with feet splayed near the yellow irises,
He smiles softly, as a convalescent child.
Let the healing woods shelter him, who is chill!

No light wind's perfume touches the soldier now,
Drowsing still in the sun, arms lax across his chest
For there are two bullet wounds in his right ride.

Seamus Heaney

Seamus Heaney was born in County Derry in 1939. His books of poems are *Death of a Naturalist* (1966*), Door into the Dark* (1969), *Wintering Out* (1972), *North* (1975), *Field Work* (1979), *Station Island* (1984), *The Haw Lantern* (1987), *Seeing Things* (1993), and *The Spirit Level* (1996). He has also published three collections of essays, most recently *The Redress of Poetry* (1995) and a version from the Irish, *Sweeney Astray* (1983). In 1995 he was awarded the Nobel Prize for Literature. He is a member of Aosdána.

News of the Raven

(adapted from *Beowulf*)

He told the truth
and did not balk, the rider who bore
news to the cliff top. He addressed them all:
'Now the people's pride and love,
the lord of the Geats, is laid on his deathbed.
Now war will overwhelm our nation,
soon it will be known, far and wide
that Beowulf is gone. They will cross our borders
and attack in force when they find out.

It is time to take a last look
and launch our lord, the lavisher of rings,
on his funeral road. His royal pyre
will melt no small amount of gold:
heaped there in the hoard, it came at heavy cost;
and that pile of rings he paid for in the end
with his own life will go up with the flame,
be furled in fire: treasure no follower

will wear in his memory, nor lovely woman
link and attach as a torque around her neck –
but often, repeatedly, in the path of exile
they shall walk bereft, bowed under woe,
now their leader's laugh is silenced,
his spirit quenched. Many a spear,
dawn-cold to the touch, will be taken down
and waved on high; the swept harp
won't waken warriors, but raven will wing it
over doomed men with his dark news,
tidings for the eagle of how he hoked and ate,
how he and the wolf made short work of the dead.'

Tony Hickey

Tony Hickey is a writer, born 1937, in Newbridge, Co Kildare. He was a founder member of The Children's Press, the Irish Children's Book Trust, and the Irish Writers' Union, of which he has been treasurer and chairperson. He has served on the board of the Irish Writers' Centre in Dublin. He is also a member of Amnesty International. His father, Captain Patrick Hickey of the Royal Artillery, was killed in action in Italy in 1944.

Sunday, September 1944

On the day my father died
My mother cried,
Collapsing into supportive arms,
'Dead?
He can't be dead!
Inside my head

There are too many
Unasked questions.'

But Grandma said,
'Of course he's dead.
I've read the wire
From the King of England.'

Kevin Higgins

Kevin Higgins is a writer based in Galway. His poems have featured in several collections and have been broadcast on RTÉ. An anthology of his poems, *The Boy With No Face*, is forthcoming.

Talking with the Cat about World Domination the Day George W. Bush almost Choked on a Pretzel

Now that pretzel's gone and done
something an expert like you never would
– loosening its hold a split-second too soon –
I think it's time we revised our strategy.
Just sitting back waiting for the big collapse?
Face facts. It isn't happening.
If there's a job to be done, why not us?

This time tomorrow, we'll be in Washington
telling Bush to come out with his hands up.
Faced with me and you, Puss, I bet he'll just crumble.
And we'll whisk him off to Guantanamo Bay
where he'll share a cage with the Emir of Kuwait.

I see from the frown wrinkling your brow,
you're worried, perhaps, how
Mariah Carey fans everywhere might react.
Too late for all that. To put it in terms
I think you'll understand: after the years wasted
here in this litter-tray, it's time to deliver
for me and you, Puss. Our battle-cry?
Something snappy? Like?
Yes, I have it! Repeat after me:
Don't make me angry, Mr. Magee.
You wouldn't like me when I'm angry.

Michael D. Higgins

Michael D. Higgins is a poet and a politician, currently Labour Party Spokesperson for Foreign Affairs. His publications include *The Betrayal*, with artwork by Mick Mulcahy (1990) and *The Season of Fire* (1993). 'Exiles' is from his forthcoming collection *An Arid Season*.

Exiles

No it is not the end of history.
Nor is it a possibility exhausted,
Not yet the end of ideas.
It is the time of a single idea,
Crippling, vicious and deadly,
Closing off from what we imagined
Of a world
We have not yet managed to create,
Rejecting the possibility,

All hope,
Of a better version of ourselves.

And in the new intolerance
We may not speak of prophecy.
We may not make a criticism
Of the choices made
In our name.
The mind of war is being remade,
New demons invented,
And language gives way
To suggestions of evil.
A picture is being drawn
Of those less than human who differ.

We are in exile.
Everywhere the spirit cries out
For a different version of our world.
An old vision of freedom
From hunger, fear, abuse,
Has faded in the terrible times.
We are invited to forget that promise
That ours was a world to create.

Out of the depths we cry
As our hearts are turning to stone.
We shrink in fear.
Few break the silence.
We accept a terrible inevitable
Prescribed, unnecessary, false, distorted.

We must make our own answer.
Our liberation from the nightmare will come.
Our exile will end,
Not from the making of miracles
But from the strength of will and heart combined,
Affirming,
That we make our own history with heart and head.
We make our common fate.
Together we move on and recall
An old promise,
Not rejected but unfulfilled.

Fred Johnston

Fred Johnston was born in Belfast and currently lives in Galway where he is Director of the Western Writers' Centre. His collections of poems include *A Scare Light* (1985), and *Song at the Edge of the World* (1988). *Powers Without Maps* (2003) is his most recent publication.

No War Then

To The Lighthouse lay on a pillow
Big enough for both of us.
The curtained room was warm, quiet –

We made love here. No war then.
The radio was a long way off,
A voice in another part of the house.

A gasometer gloomed on the garden,
Blood-rust coloured; we were near
The sea, and we had a few friends.

Innocent as dust, as leaves falling –
We know better now. Too grown for
Our own good, war is everywhere.
These mad days I think (forgive me)
That it could be no possible sin now
To feel your breath in my breath
In such a warm, quiet room.

Jennifer Johnston

Jennifer Johnston is a novelist and playwright living in Derry. Her novels include *The Captains and the Kings* (1972), *Fool's Sanctuary* (1987) and *The Illusionist* (1995). This excerpt is from the novel *How Many Miles to Babylon?* published by Hamish Hamilton (1981).

from *How Many Miles to Babylon?*

A thaw set in and the earth, brown again for a while, sucked at our feet as we marched through the countryside. The men were in low spirits, slow to obey orders. Endless troop transports churning mud pushed us into the ditches, breaking the formality of our ranks, covering the men with filth. They could no longer even be bothered to shout abuse.

Across the grey sky from south to north came two swans. They were flying low, their wings fanning with dignity the air around. I stopped marching, embarrassed by their presence, as if some old acquaintances had dropped in to visit me at an unbearably inconvenient moment. They skimmed the bare branches of six or seven battered trees and flew obliquely across the line of soldiers. As I raised my hand in greeting the sound of a shot reached me. The front bird's neck swung for a moment from left to right and then drooped. An ugly mass of flesh and feathers fell to the ground. The men broke ranks and ran to look. The living bird faltered for the moment and then flew on, adjusting its flight upwards towards the safety of the clouds.

'Who did that?' My voice was blown back at my own face by the wind.

They pushed each other to see the bird. One of its wings was broken by the fall and was crumpled under the heavy body.

'A great wee shot, man. I never knew you were that handy with a gun.'

'May God protect the Kaiser if you're ever around.'

'Who the bloody hell did it?'

'I did, sir.' A small man waved his gun cheerfully at me.

'Just why?'

All the muscles in my face were trembling.

He shrugged, dismissing me and the dead swan simultaneously.

'Where's the harm?' asked someone.

I turned and walked away.

Someone made an obscene noise and then the crisp voice of the N.C.O. ordered them back into their line again. They sang all the way back to the farm.

Pat Jourdan

Pat Jourdan lives in Galway. Her poems have appeared in *Rialto, Aquarius, Strictly Private, Cúirt* journal, *Burning Bush, West 47, Orbis* and broadcast on BBC radio 4's 'Poetry Please', RTÉ's 'Sunday Miscellany', Radio Norfolk and Flirt FM. Her most recent collection is *The Bedsit* (2002).

After War

Country lane, August, just after rain;
that washed-leaves, rinsed-grass smell
and the chugging of an approaching lorry.
Passing us by, its tailgate rattles
and several blond youths lean out
waving at us in late afternoon gold.
Such enthusiasm, I wave brightly
and meet the eyes of one young man,
his extra-vivid smile for me alone.
Childhood sunlight dashes between us;
grandmother's hand snatches my arm,
'Prisoners,' she hisses, 'Prisoners-of-war. Don't wave.'
Because of them we dashed through air-raids.
Because of them our streets were on fire.
His eyes I can remember to this instant
and the idea of becoming a rebel,
being on his side –
his golden hair, the blue eyes, the sunlight,
that country lane – all the ingredients of love,
only war keeping us apart
and the rules of adults' time and enmity
on that lane that wandered to the shore

and the clanking chain
that kept the prisoners secure
as the lorry blundered off into a quiet distance.

Brendan Kennelly

Brendan Kennelly was born in Kerry in 1936. A poet, dramatist and critic, he is currently Professor of Modern Literature at Trinity College, Dublin. He has published many collections of poetry including: *My Dark Fathers* (1964), *A Small Light* (1979), *The Boats Are Home* (1980*), Moloney Up and At It* (1984), *A Time for Voices: Selected Poems 1960-1990*. His dramatisations include *Antigone* (1986) and *Medea* (1988). 'A Holy War' is from *Cromwell* published by Beaver Row (1983) and Bloodaxe Books (1987).

A Holy War

from *Cromwell*

'We suffered the little children to be cut out of women
"Their bellys were rippitt upp"
This was a holy war, a just rebellion
And little lords in the womb must not escape
Their due. Certain women not great with child
Were stripped and made to dig a hole
Big enough to contain them all.
We buried these women alive
And covered them with rubbish, earth and stones.
Some who were not properly smothered
Yet could not rise
(They tried hard) got for their pains
Our pykes in their breasts. People heard
(Or said they heard) the ground make women's cries.'

Susan Knight

Susan Knight is the author of two novels, *The Invisible Woman* (1993) and *Grimaldi's Garden* (1995), and is the editor and compiler of a collection of interviews with immigrant women in Ireland, *Where the Grass is Greener* (2001). Two sections from this were included in the *Field Day Anthology of Irish Writing Vol. 5* (2002). In addition, she has written for stage, radio and television. 'The Meadow' first appeared in 1998 in *The Stinging Fly* and is reproduced by kind permission of the author and the editor.

The Meadow

I am standing in the meadow watching him walk away. I am standing in the long grasses of the wide meadow. I want to call out to him. I want to will him to turn and look at me but he just continues to walk away, further and further.

The grasses are pale: bleached yellow, bloodless pink, dry green. They wave their seed heads, brushing my thighs. They undulate like the surface of the sea.

Turn, I call silently. Turn back to me. He walks away, along the straight path beside the meadow. I can see him seem to get smaller and smaller for there is no twist in the path. I can watch until he seems to disappear altogether, realising that I too will have disappeared. If he turns now he will see that I too have almost disappeared.

And what if I call out loud? What if he turns and waves goodbye? What if he turns and still fails to stop? How will I live with that?

If he turns now and sees the great sea of the meadow, its waves swirling and churning, perhaps he will think I have drowned. Perhaps he will think I was never there. That the form he thought was me, so carefully not looking, was just a scarecrow, just a sack stuffed with straw.

I'm a plain woman. I can't pretend otherwise. I can't pretend there was ever even in the nape of my neck a sweetness to draw a man back from an action he'd decided to pursue. My body was never that of a young girl, but always thick, my flesh coarse, my hair dry as straw. That's why I don't call out. Because I don't think it would make him stay. At least as things rest the answer to the question remains open. I stand in the meadow always – my arms stretched out to him as he walks away, his name screaming silently out of my throat.

That was long ago. And while my real self stands still in that eternal moment amid the long meadow grasses, my shadow moves through a series of automatic gestures in what people are pleased to call time. I marry a man from the village, a large, comfortable man who appreciates my aptitude for hard work despite my lack of feminine graces. Between the weary grind of work and the blessed absences of sleep we make five children. Then I curl up against his back, the two of us fitting together like pieces of a jigsaw puzzle. And when I place my hard hand on his soft, hairy belly, this is the best moment of all. This and those spent in a half-dream suckling the five babies, one after the other, so drunk on the sweet smell of their sweat that I almost forget that I'm not really there. That I am still standing in the wide meadow watching a young man swing away from me down a straight path of hard clay, without turning, the wind that lifts the grasses lifting his long dark hair.

Word reaches our village from time to time from the outside world. Word of wars fought and lost – for wars are always lost: how can even one death in battle be termed a victory? Word of uprisings and revolutions, cataclysmic accidents, acts of a God we are told loves us. Little touches the grind and sleep, grind and sleep of our lives. Word comes of him too but I laugh. I know his real self has never gone but is still eternally going. That the one who made it to the city is just a shadow, like me, like the me that laughs at the news of him.

Some of my children marry and have children of their own. One goes to the city and disappears there, one goes further and is killed in a battle some call a victory. My man dies suddenly in his bed and I awake beside a cold and stiffening corpse. I miss the comfort of him at night and I weep for it.

I am standing in the meadow. I am standing in the long grasses of the wide meadow watching him turn back to me. He is running towards me, his arms outstretched. They have told me that it is his son, so like him as he once was you can't tell the difference. But I know better. They say he died in the city and that his son is bringing his ashes home, to scatter on the meadow of grasses. It was his wish, they say. To bury his heart where he left his heart, where he would have stayed if the woman he loved had said just one word. If she had not indifferently watched him walk away from her. That's what they say the son said.

I am standing in the long grasses. The wind blows against me like kisses. He has turned towards my outstretched arms and is running back towards me. Always. Forever.

Conor Kostick

Conor Kostick is an historian and writer based at Trinity College Dublin. His publications include *Revolution in Ireland* (1996) and, together with Lorcan Collins, *The Easter Rising – A Guide to Dublin in 1916* (2000). He is currently Chairperson of the Irish Writers' Union and a reviewer for the *Journal of Music in Ireland*.

When George Bush referred to the need for a 'crusade' against terrorism, he was using the term loosely, in the sense of a vigorous campaign. But he, or his advisors, had chosen a word heavily laden with meaning, a term which in the Middle East has very negative connotations.

The medieval crusades shocked the Arab world, particularly in the ferocity with which they dealt with the citizens of captured towns. The fall of Jerusalem in 1099 was an especially bloody affair and became a cause célèbre in the later emergence of Arab nationalism.

A Provençal priest, Raymond D'Aguilers, was present when the First Crusade captured Jerusalem and wrote an extraordinarily vivid account of the event. His work has many fascinating details, but it also stands as evidence that when the Bush administration evokes the spirit of the crusades they are unconsciously associating themselves with the horrific slaughter of civilians, carried out by Christian people who saw themselves as thereby glorifying God.

I have translated the following excerpt from Raymond's Historia Francorum qui ceperunt Iherusalem *for this anthology.*

The Fall of Jerusalem, 1099.

[*14 July 1099*] The sun rose on the day we had appointed for battle and our all-out assault began. But firstly I wish to say that according to my reckoning, and that of many others, there were as many as 60,000 Saracen soldiers inside Jerusalem, not counting the innumerable women and small children.

What of the strength of our forces? I calculate that the total did not exceed twelve thousand, excluding the many ineffectual and poor people and out of this army I reckon that no more than 1,200 or 1,300 were knights. I have told you this so that you may understand that whether great or small, nothing that is undertaken in the name of the Lord is in vain, as the next page proves.

So, our army began to haul our siege towers to the city walls; stones cast from windlasses and catapults were flying on all sides and arrows surrounded us like an immense hailstorm. The servants of God endured this barrage, conducting ourselves with faith whether meeting with death or obtaining vengeance upon the enemy. At this time the battle gave no indication of victory, but now we drew near to the city walls with the siege engines.

Not only stones and arrows but also firewood and straw were being thrown down onto our machines and on top of these were hurled flaming mallets – made from wrapping them in pitch, wax, sulphur, tow, and rags which had been set on fire. These mallets, I should explain, having been covered all over with nails, were intended to stick fast upon striking our machines, in order to set them on fire. At all events they were able to hold us back more effectively by the fire that was kindled on the wood and straw than with their swords, fortifications and deep ditches.

Battle was conducted from the rising of the sun to the finish of this miraculous day; on no previous occasion have such extraordinary deeds taken place. All the while we were appealing to God, the omnipotent commander and leader of our people, trusting to his mercy.

The sun was setting and as darkness fell dread came over both sides. For their part, the Saracens were terrified that with their outworks being broken and the ditch filled the interior wall of the city would be breached during the night or with the coming of day. On the other hand our people feared only this: that the Saracens might somehow be able to set fire to our siege engines and thus enormously strengthen their cause. Therefore on both sides a night of patrols, toil and sleepless anxiety

As the night passed a most definite confidence arose among us, while a growing dread sapped their courage. For while we were trying to seize the city for God by an act of free will, their support for the laws of Mohammed was based on coercion. All the same, both sides were making extraordinary efforts throughout the night.

[*15 July 1099*] As the morning dawned high spirits lay upon us with such intensity that we attacked right up to the city walls and brought our machines with us. But the Saracens had made so many siege engines that nine or ten of theirs were opposing each one of ours and as a result they checked our vehement attack. A rain of stones was battering our machines and with our forces tiring there were some who began to lose heart. However, the invulnerable and constant mercy of God remained with us.

I cannot pass over this pleasing incident. When two women were about to pluck out the stone from a windlass by bewitchment, the same stone was powerfully cast out and it crushed them together with three very young girls, thus smashing out their souls and turning aside their enchantment.

Around midday all our people were becoming very demoralised, as much from exhaustion as from despair. This was understandable seeing that we were opposing an enemy with solid and tall city walls, with great numbers, and with an advantage in the way in which they could deploy their defences to face us. Although our people were faltering in this way and the exultation of the enemy was growing, nothing could come between us and the compassion of God, through which our lamentation was converted to joy.

For just at a time when a counsel was being urged to retreat by certain men, since some our siege engines were reduced to ashes and others were shattered, a knight on the Mount of Olives began to brandish his shield, signalling to the followers of Count Raymond of Toulouse and to others that they should advance.

Who that knight was remains a mystery to me. Inspired by this signal, our people, who had been listless, were greatly invigorated; some began to attack the walls while others put up tall ladders and hurled long ropes.

Moreover, a certain young lad had devised arrows covered with burning cotton which he fired against the rampart of the Saracens that faced the wooden tower of Duke Godfrey of Lotharingia and Count Eustace of Boulogne. These arrows kindled a fire that drove back those who were fending off our tower.

Next the duke and those with him swiftly lowered the wickerwork that had previously protected them; it was swung from the middle of the tower to make a bridge onto the walls. Bravely and fearlessly the duke and his men began to penetrate Jerusalem.

Among the first to enter were Tancred and Godfrey, who shed so much blood that day, it is unbelievable. Everyone was climbing up after them and now the Saracens were really suffering. I should mention something remarkable, that even though the people of the city were nearly in the hands of the Franks, there was still a group of Saracens facing Count Raymond who fought as if they would never be taken. But with our people now on the walls and towers, we had become masters of the city.

Thereupon the spectacle was wonderful. Some of the citizens were simply decapitated – which was the lightest fate; others meanwhile were being compelled by our arrows to jump from towers; yet others, in truth, were twisting all day to flames before being consumed by fire. Heaps of heads, hands and feet of the citizens could be seen piled up along the rows of houses and streets. Truly, our knights and footsoldiers were dashing back and forth over the dead men and women.

All that I have said thus far is unimportant and trivial compared to what happened at the Temple of Solomon, where the Saracens had become accustomed to chanting rites and performing rituals. What happened there? If I speak the truth I go beyond belief. Let it merely suffice to say that in the Temple and its colonnades our knights *rode in blood to the knees and bridles of our horses.*

It is no doubt a proper judgement that the very same grounds on which blasphemy towards God was endured for a long time received the blood that was spilled. Jerusalem was thus filled with corpses and the blood of the citizens.

This day will be famous forever.

Morgan Llywelyn

Morgan Llywelyn was born in New York city to Irish parents, but now lives and works in Dublin. She writes for adults and for young readers. Her publications for adults include *Lion of Ireland* (1980), *Druids* (1991), *1921: a Novel of the Irish Civil War* (2001), and, for children, Bisto award winners *Brian Boru* (1990) and *Strongbow* (1992), both published by The O'Brien Press. She is a founder member of the Irish Writers' Centre and a former chairwoman of the Irish Writers' Union. The extract reproduced here is from *1921*, published by Forge, New York, a subsidiary of St. Martin's Press, and is reprinted by permission of the author.

The events described here are based on the report of Commandant Paddy Ó Dálaigh (O'Daly), Military Archives, Dublin, and the filmed personal testimony of Stephen Fuller, shown in a documentary, Ballyseedy, by RTÉ.

from *1921*

Paddy O'Daly looked up from behind a desk cluttered with papers, boxes, bits of military paraphernalia, and the dried-out remains of an uneaten sandwich. Henry remembered O'Daly from his days as one of the murderous Twelve Apostles. Then he had been a deceptively gentle-looking 'civilian' with a disarming smile and a thick crop of curly hair. Now he was attired in full Free State uniform and an officer's cap crushed his curls. There was no smile.

'Oh it's you, Mooney. Heard you were in Castleisland.'

'Yesterday,' Henry replied.

'Yes. So you know.'

'I know what happened there. What's happened here?'

O'Daly shifted in his chair. When he spoke his voice was a rapid staccato, as if repeating words memorised by rote. 'Immediately following my notification of the atrocity at Knocknagoshel, I issued an order that in future government forces are not to remove any barricades, clear any dumps or touch any suspected mines. Irregular prisoners are to be fetched from the nearest detention barracks to remove them. This is a necessary precaution to save the lives of our own soldiers, and within hours has proved its value.'

O'Daly lifted a sheet of paper from his desk. 'Here is a statement for publication.'

Henry read, 'While a party of government troops was proceeding from Tralee to Killorglin on the evening of March sixth, they encountered a barricade of stones at Ballyseedy Bridge. The party returned to Tralee and brought out nine Irregulars who were instructed to remove the barricade. While they were engaged in doing so at about two a.m. on March seventh, a trigger mine exploded and the prisoners were killed. Three members of the Free State Army suffered shrapnel wounds.'

Henry felt sick to his stomach. 'Prisoners from Ballymullen.'

O'Daly gave a terse nod.

'May I have their names?'

'They won't be released until all the next of kin have been notified.'

'Was one of them Ned Halloran?'

'Can't tell you that. It may take some time to identify them. A bomb hidden among stones does a lot of damage.'

Henry stared at O'Daly. 'Nine IRA men at two in the morning. The exact time and the same number of men as at Knocknagoshel.'

O'Daly stared back at him. 'Coincidence,' he said.

Ballyseedy Cross was easy to find. Henry had passed by the place many times on his bicycle. But never had he seen the sight that greeted his eyes that evening. A great hole had been gouged in the road. Huge splashes of blood; gobbets of unrecognisable flesh, naked and vulnerable as meat in a butcher's window. Everywhere. On the crushed grass, in the dripping trees. The air stank of it. And of explosives. And hatred. And death.

Dear Jesus, Ned, was this how it ended for you?

Birds were picking at the shredded flesh. A few local people appeared, tentatively, at the edge of the woods, or farther down the road, and stared. Eyes wide. Crossing themselves repeatedly. Crying, some of them. One man had a pony and trap. The pony shied violently at the smell of the blood and would come nowhere near but stood at a distance, trembling.

A cap trampled in the mud. A broken rosary. Part of a sleeve, the buttons still bright, the hand ...

'No!' Henry cried, the word torn out of him, wrenched from his guts by grief and horror. 'No! They are us! They are *us*!!!'

He got off his bicycle and vomited into a ditch. Then he made his way to the stream spanned by the nearby bridge and washed his face. There was blood there too, a sticky puddle in the mud on the bank, but he was careful not to touch it. Could not bear to touch it. Its silent voice cried out to him.

He knelt to pray.

I don't even know what to say to you. Are you blind? Deaf? Dead? What sort of deity are you, to allow this?

'Why bother?' he asked himself aloud. He stood up and brushed off his knees. 'That's me done. I'm through with God.' *You hear me? Through with You!*

Re-mounting his bicycle, Henry returned to Tralee. He wobbled as he rode. He was trying to hold the pain at arm's length but if his concentration lapsed to the smallest degree it rushed in on him.

By dawn next morning he was back at Ballymullen, trying to get a list of the dead. There was still none available. The guard remembered him from the day before and let him in anyway, thinking O'Daly had sent for him.

Henry pretended to take a wrong turn and wandered through the building. The barracks was depressingly grim. Cell-like rooms, narrow passageways, flaking paint, insufficient light. As he passed an open doorway Henry noticed a sandy-haired Republican he had known before civil war broke out. A man who was now in the Free State Army. He was standing behind a table, sorting out personal effects taken from the pockets of dead men.

Henry entered the room and struck up a conversation, being careful not to look directly at the objects on the table. Slowly he brought the topic around to them. 'The bombs at Ballyseedy did so much damage it's hard to identify the men.' the soldier told Henry. 'All the remains of one of them could be fitted into his tunic with the sleeves knotted. But there is something … wait a minute, I'll show you.'

Bombs? Bombs plural?

The soldier produced a slip of paper with a badly charred edge. The printing on the paper was almost illegible with blood, but when Henry held it up to the light and squinted, he could make out part of a safe conduct pass signed by Michael Brennan.

Blood dripping from the trees. Birds picking at ruined flesh.

He could not go to bed because he did not dare close his eyes. The nightmare would get him if he did. His brain sat thick and heavy in his skull, but he could not make it work. No moving parts. Just a dense, soggy organ, inert. He knew he should think but he did not remember how to think.

Exhausted, he wandered the narrow lanes of Tralee and cried out in his heart for Ella, for Ursula. For somebody.

Other Republican prisoners were taken from their cells and transported to a barricade at Countess Bridge. Bombs in a barricade there killed four of them. Five more died in similar circumstances near Cahirciveen.

Horror upon horror. Hatred expanding outward in a great wave.

On the ninth of March Paddy O'Daly sent a message to Dublin. 'It has now transpired that an Irregular prisoner, Stephen Fuller, escaped during the mine explosion at Ballyseedy. It is stated that he has become insane.' A reward was issued for

Fuller's immediate capture.

The mutilated bodies of eight men had been distributed among nine coffins.

Henry recognised the comment about Fuller's 'insanity' as an attempt to discredit any version of the event the man might tell. But terrible stories were already circulating through Tralee and the surrounding countryside.

To Henry's frustration, the names of the dead were still not being made public.

Recalling that someone else had carried a safe conduct pass from Michael Brennan, Henry went back into the hills, to the little cabin with the swaybacked roof.

He arrived to find a family in mourning. They had just been notified by a government messenger of the death of their patriarch. At Ballyseedy.

As if they heard a banshee wail on the wind, the neighbours were already gathering.

Knowing Ned might still be alive did not spare Henry from grief. No one in that cabin was spared; grief came to meet each person who entered, wrapped its cold arms around them, opened its gaping maw and swallowed them whole.

'Sorry for your trouble,' Henry whispered to the widow, repeating the time-honoured phrase. Then he went to stand with the other men beside the hearth. Passing around the whiskey. Smoking clay pipes. Talking in low voices.

After a time the door opened and three figures entered, two men and someone who was wrapped in a shawl like a woman, but did not move like a woman. He edged crabwise, like a man in great pain.

Curious, Henry drifted closer to hear this person say to the widow, 'Sorry for your trouble, missus. My name's Stephen

Fuller. I knew your husband well; I was with him when it happened. I come to tell you he was brave to the end. God took him without letting him suffer none.'

The woman lifted trembling hands and put them on either side of his face, pushing back the shawl to reveal a gaunt young man with boulders for cheekbones, and haunted eyes. His face was badly bruised.

'You were with my man at Ballyseedy?'

'I was.'

Suddenly her legs gave way. She collapsed as if her bones had turned to chalk dust. While the other women rushed to help her, Henry moved closer to Fuller. 'Would you be willing to talk to me?' he asked. 'I used to write for *The Irish Bulletin*.'

Fuller's eyes scanned his face. 'No word of a lie? I used to read the *Bulletin*.'

'Then you know my name: Henry Mooney. Can you tell me what really happened at Ballyseedy?'

'You can't write about it, not now,' Fuller said.

'I realise that. But a friend of mine may have been there; I have to know. Ned Halloran?'

'There was no Halloran with us. It's one of my own friends we're mourning today. Them who brought me here didn't want to but I owed it to my friend. I'll be leaving soon, though. The IRA has a safe house for me back in the hills, and in a few days there'll be a dugout ready for me to hide in until they can get me out of Kerry. You didn't see me at all, remember.'

'You have my word on it.'

Fuller whispered urgently, 'Don't write this down. It could mean your life as well as mine. Just listen.

'I was in a cell in Ballymullen the evening the government soldiers came for us. They told us, "We're going to blow you up with a mine." We were removed out into the yard and loaded into lorries, and made to lie flat in the bottom and taken out to Ballyseedy crossroads. We couldn't see nothing and they told us nothing, but we knew it was bad.

'We arrived out anyway and they marched us up to this pile of stones and logs in the road. A barricade, like, but you could see it wasn't no real barricade. The language the Staters used to us was fierce. One fellow called us Irish bastards and he an Irish man himself. One of our lads asked to say his prayers and they told him, "No prayers. Our fellows didn't get any time to say prayers. Maybe some of you will go to heaven and meet some of our fellows there."

'They tied us then with our hands behind us, and made us stand in a circle with our backs to the stones. And they tied our legs and tied our shoelaces together too so we couldn't run. The lad next to me closed his eyes and started praying but I kept watching, like. The lad on the other side said goodbye and I said goodbye, goodbye lads. And I saw them pull the trip-wire. And up it went. And I went up with it.'

Henry realised he had been holding his breath. 'Government soldiers set off the mine themselves?'

'They did surely. More than one, I think. There was a boom and another boom, and maybe another again.'

'How in God's name did you get away?'

'When I saw them pull the trip-wire I threw myself sideways, like. Next thing I knew, I was lying in a ditch with the ropes blown off me. Me back and hands and legs was burnt something terrible. Our lads was writhing in the dirt with stones raining down on them and guns was blazing, the Staters was shooting them on the ground like dogs, and everything was burning and screaming and …' Fuller's words ground to an agonised halt.

Tears streamed down his ravaged cheeks.

Henry gave him a cup of whiskey. After draining it in one long swallow, Fuller continued. 'In all that confusion no one noticed me. I crawled into the stream, right down under the surface in the muck, and escaped that way. I don't know how long I was dragging meself through the night and across the fields before I found help. It's all of a mist now. A red mist.'

'Were any of the government soldiers hurt?'

'The bombs didn't take a feather out of them. They took care to stand well away while they blew us to pieces.' Fuller drew a ragged breath and fixed his eyes once more on Henry's face. 'What happened?' he asked in baffled tones. 'All us lads who fought side by side to make Ireland free ... what happened to us?'

'I can tell you,' said Henry, 'because I've chronicled it day by day. But I don't think you or I will ever understand it.'

Sam McAughtry

Sam McAughtry was born in Belfast in 1921. His novels include: *The Sinking of the Kenbane Head*, (1977), *Touch And Go* (1993). Short-story collections: *Play It Again, Sam* (1978), *Blind Spot and Other Stories* (1979), *Ward & River* (1981). His most recent collection of short stories is *Hillman Street High Roller: Tales from a Belfast Boyhood* (1994). *The Sinking of the Kenbane Head* was published by Blackstaff Press in 1977.

from *The Sinking of the Kenbane Head*

Mart, 1935.

'Mart went to sea as a trimmer, in a Head Line ship, shovelling coal down a chute to the firemen in the stokehold, when he was eighteen and I was eleven. When I left him down to the ship and said goodbye to him in the dark fo'csle where he would live from then on during his voyages across the North Atlantic, he shook my hand. It was the first time that anybody had ever shaken my hand in earnest, and it was the first time I had said "Bon Voyage".

'When I came back home I went upstairs, shut the door, lay down on the bed, and cried like a baby. He was my great friend and defender and I loved him fiercely. Mart protected me when Jack, the oldest, cuffed me. He listened to the stories that poured out of me on our walks over the Cave Hill and he pretended to believe them all. He talked to me as though I was a grown-up, instead of a run-down, glandular kid, who had very nearly joined Mary, Molly, Betty and Harry in their tiny graves at Carnmoney. When he sailed away I wrote long letters to him and he wrote back from places like St John's New Brunswick, Quebec, Boston, and Philadelphia, telling me how much he enjoyed my stories.

'When he came back from that first trip, I ran down the street to meet him. Ten yards away I stopped. He looked like death, his eyes were sunken, his face like a skull. Mother bit her lip at the sight of him: "Was it bad, son?" He said that he was all right now that he had his sea legs, then he gave her a big hug and winked at me over her shoulder with his poor wreck of a face.

'As I sat beside him on the sofa I took one of his hands and turned it palm-up: the whole surface was calloused and pitted with specks of coal in a hundred places. He smiled at me: "I've got real sailors' hands," he said. That was one of the things that the Head Line Shipping Co. Ltd gave away along with the wage of £9 a month – real sailors' hands.'

* * *

The Start of It, 1939.

'When Neville Chamberlain announced on the radio that Great Britain was at war with Germany my mother slid back in her chair and fainted. Since our sister wasn't there at the time I ran next door to Mrs McClean for help. Kindly Mrs McClean brought Mother around: "I know why you're upset," she said, "so go ahead and cry." And Mother cried because her brother Billy died at the Somme in 1916, and because her man was a merchant seaman, sailing the North Atlantic with my brother Mart, and because my brother Tommy had told her that he was going to join the North Irish Horse and I had told her that I was going away in the RAF as soon as I could manage it. Four of us would be fighting on land, sea and in the air. Mother had plenty to cry about on that Sunday morning.'

* * *

Mart's Death, November 1940.

'The scene from on board was one to strike fear into the heart of any seaman. The crew on deck had seen the whole horizon to their starboard quarter explode in one massive eruption as the *Maidan* blew up. Suddenly a star shell burst overhead, bathing the freighter in cruel blue-white light. The *Kenbane Head* was outlined against the night backdrop as though she was the only ship in the sea and the ocean itself had frozen. The tired little ship seemed to pause, immobile for an instant of time, turning her throat to receive the death blow from the powerful, invisible enemy poised outside, in the darkness.'

* * *

August, 1944.

'On 31st August 1944, over the Aegean Sea, I was navigator in one of six Beaufighter anti-shipping aircraft. We each carried eight rockets, four 20mm Hispano cannons, six machine guns, and I had a Browning machine gun to myself in the rear cockpit. Suddenly someone called. A sighting. A destroyer with an armed escort, both German. I could see fighters. I went to the radar screen; "2,000 yards," I called, "1500 yards – hey, Johnny?"

'My pilot, John Bates, DFC said, "Yes, Paddy?"

'"Hurt these bloody bastards," I said, "eight hundred yards – Fire!" And eight 60lb rockets went howling down like banshees towards the grey, lean target below. We arrowed on to the ship, cannons yammering as we strafed the deck, going for the bridge and the gunners.

'At just above mast height we banked and jinked away fifty feet above the waves as I opened up with my single gun, spraying the smoking deck hopefully.'

* * *

January, 1946. After the War.

'Mart is entitled to his special epitaph. He won't mind sharing one used often by the Germans who killed him, and who died themselves, in turn:

> '*Auf einem Seemannsgrab da bluhen keine Rosen ...*
> There are no roses on a sailor's grave,
> No lilies on an ocean wave;
> The only tribute is the seagull's sweeps.
> And the teardrops that a loved one weeps.'

Eugene McCabe

Eugene McCabe was born in Glasgow in 1930. He has lived on the Monaghan/Fermanagh border since 1955 where he was a fulltime farmer, part-time writer. Now retired, he writes occasional verse. His publications include *Heritage and Other Stories* (1978), *King of the Castle* (1978), and *Death and Nightingales* (1993).

Everything Changes, Nothing Changes

This is an extract from the jubilee keynote address given to the Clogher Historical Society, November 2002.

Everything changes, nothing changes, essentially.

Through my work I have tried to explain, explore and exorcise the violent events in adjacent towns and neighbouring town lands. It's clear that we all participate and are all survivors and/or victims of history. That's the thread running through almost everything I've written; local stories elaborated and thinly disguised, all of them attempting an explanation of what seems almost inexplicable to the outside world.

The outside world arrived once in the seventies in the form of a Swiss TV camera crew. I was put standing against the blasted bridge at Lackey with its half-circle of spiked concrete bollards daubed with predictable slogans. The British army were watching from a high lookout about three hundred yards away. I answered obvious questions with obvious answers, and we went back up to the house for coffee. At some point during this refreshment the director said, in flawless, accentless English which they almost all have: 'Of course, Ulster is the yawn of Europe, it's four hundred years out of date.'

This was said not as a gibe but as a boring matter of fact. I'm not sure why I felt both defensive and offended. He saw what Churchill called 'The integrity of their unchanging quarrel' as a wretched, murderous little scrabble in a remote corner of Europe. Naturally, as his host I did not respond with a mutter about Swiss banks refusing to part with the vast, confiscated funds of incinerated Jews *without death certificates* – which the Nazis were not in the habit of issuing. Better a yawn at the edge of Europe, I thought, than a disgrace in the middle of it!

* * *

About three miles or so from where we live there's a lake called Lough Inver. It has a stony edge, a long lake lined with mature beech trees. The branches grow down into the water. It's very beautiful. When we first took this walk many years ago we stopped to admire an old house, very typically Fermanagh – three pointed dormer windows, the middle one over the hall porch, all glass, like a small, upstairs conservatory. It looked late-eighteenth or maybe early-nineteenth century. There were mature lime trees going up a small avenue, a profusion of daffodils and bluebells in season. A few hundred yards from the house there was a small church or hall by the lakeside with an inscription above the door on a plaque or shield which read:

> Inver Hall.
> AD 1868
> And underneath that;
> 1690

Clearly, we were in planter territory. There was evidence of order in the fields, the houses and the stone-built outhouses. Years passed. Every now and again we would go for the Inver walk and each time we noticed that the house with the dormer windows looked more desolate, till gradually it became so overgrown you could hardly see the house from the road. And yet there was smoke coming from the chimney. Someone was living there. Obviously something had happened, but it seemed almost more attractive in neglect than when orderly and cared for. Naturally, the imagination goes into overdrive.

Some strange story. You think of Miss Havisham in Dickens's *Great Expectations*. The whole place emanated melancholy in a way that was palpable. We were so intrigued I was almost tempted to go up on some pretext. I wanted to put a human face on this gradual dissolution.

I thought I was fairly familiar with everything that had taken place in my own area (murders and bombings), the when, the where, and the how – this field, that family, that crossroad, that laneway – but this house at Inver I did not write about or refer to because I had only heard about it recently when I was talking to a Fermanagh neighbour, Robert Irwin. I knew his brother Willie had a farm in the Inver area. I described the overgrown house and asked who owned it and why it had fallen into such disrepair. He looked at me for quite a while before replying with a question: 'You must know?'

'No,' I said. 'I don't.'

'They're Crowes. The father's dead this brave while. Two daughters. One stayed home to help the mother with the farm. The other lassie, Sylvia, got a bus every day to work in Enniskillen. She was on her way to the bus stop beyant near Dernawilt when a bomb in a lorry meant for the army went off. There was nothin' left of her; they had to weight the coffin with stones.'

'Weight the coffin with stones!'

This is a long way from Miss Havisham or any flavour of romance. I found it hard to believe I could live a few miles from such dreadful misfortune and not know about it. Maybe we were elsewhere for an extended period. I'm not sure which philosopher said history is like a spiral staircase strewn with blood and broken bones, but no matter how terrible it may seem at the time, the spiral is climbing upwards and despite what we read, see and hear every day, things are, imperceptibly, getting better. Now some years into this new millennium, how many of us are prepared to buy into that theory? All are agreed that the last century was the worst by far in human history. According to the Red Cross there are over fifty wars being waged around the world at the moment, and over seventeen million refugees. The heart of darkness still resides in that most dangerous of all animals, us human beings.

* * *

What is curious during the last three decades are our differing responses to the same images and events. I can't now remember where I read a first-hand account by a participant of the Burntollet march – *The Irish Times*, I think, but I'm not certain. No doubt someone will know the piece. The writer (a student) described how RUC officers marched alongside the marchers as a buffer against a threatened sectarian encounter. This lad had become friendly with a particular officer, pausing to rest, eating and sharing sandwiches, exchanging jokes and cigarettes, but naturally avoiding any talk about the political aspect of the march. There was almost, he conveyed in the article, a bonding, a sense of camaraderie.

Then came Burntollet. As the stones began to rain down, the student looked at the officer much as to say, 'What now?' What happened was the officer's fist came straight into his mouth, breaking some of his teeth. Then the stone-throwers appeared over the hill, running and howling, carrying billhooks and hedge knives. The startling violence and terror of that detail stayed in my mind because in a way it's analogous to Frank O'Connor's story 'Guests of the Nation'. Set during the War of Independence, a few captured English Tommies become friendly with their IRA captors. Days and nights pass. The *craic* is mighty, the slagging continuous, all of it good-natured. Then word comes from on high. The hostages have to be killed. They are taken out to a bog and shot. The end of that story is devastating to read. One detail amongst thousands in a war becomes an unforgettable human document. And an indictment. That story is a case in point, where the 'truth' of fiction takes precedence, in my view, over unvarnished historical fact. It also implies that love is more powerful, more enduring than corrosive hatred. Between the lines it's saying: Why can there not be an end to historic hatreds. Are we programmed to hate and suspect each other's culture and religion to the point of killing? Forever? Greek and Turk, Arab and Jew, Serb and Croat, Catholic and Protestant, Christian and Muslim? Why, why, why? The whys are apparently unending and unanswerable.

Eamonn McCann

Eamonn McCann is a journalist, a regular columnist with the *Belfast Telegraph* and a committed socialist. He is the author of many books including, *War and an Irish Town* (1974), and *Bloody Sunday in Derry: What Really Happened?* (1992). 'Remembrance' was written for the Ulster Society and published in their 1997 anthology of the same name. The lines quoted at the end of the article are from Wilfred Owen's '*Dulce et Decorum Est*'.

Remembrance

What good reason could there be for Garvaghy Road residents to oppose the march from Drumcree, a flustered and angry Orange leader demanded to know on television last July? The point of the march was to remember those who had fallen at the Somme.

To remember them, he elaborated, 'with pride and reverence for the sacrifice they made.'

It occurred to me at the time that that was a good reason for unease, at least, at the march. Not that the dead of the Somme should not be remembered, but that they should not be remembered with 'pride and reverence'. That instead they should be remembered with rage.

When we think of the Somme, as we should every year, we should rage against those responsible for sending the young men of Ulster, and from all other corners of Ireland and Britain, out to die so uselessly, in such droves.

If we look back now at the propaganda posters and political rhetoric of the time, we see World War One promoted as a clash between civilisation and bright progress on the one hand, barbarism and backwardness on the other, between the forces of good and the forces of evil. It's the same way the ruling class always presents the wars it wants the working class to do the dying in.

Hardly anybody represents World War One as such a straightforward morality tale now.

Indeed, it is remarkable that a conflict which is commemorated with such ceremonial pomp every year, supposedly so emblazoned into our consciousness that television announcers must perforce wear the symbol of its remembrance while intoning the sports results, it is remarkable that reference is so rarely made to what it was actually *about*.

Ask at random on the streets of Belfast, Derry, Enniskillen, wherever, ask what caused World War Two and the answer will come readily enough. It was against the Nazis, to stop Hitler. The Gulf War, the Falklands/Malvinas, the Boer, the Balkans, the American Revolutionary wars – most of us could make a stab at identifying the issue at stake in each. But World War One, the 'Great War' – what was that *about*?

The reason it is so rarely discussed is that even today our rulers are uneasy about the truth.

I do not believe it would be possible now to find a reputable historian, whatever his/her personal perspective or political beliefs, to argue that either side at the Somme operated from high moral ground. The main protagonists were competing sets of robber-barons, British and German capitalisms, locked into a life-or-death struggle for protected markets and access to cheap raw materials around the globe.

They fought for the 'right' to rule the waves and rob the world. Or rather, they didn't. They didn't fight at all. They never do. They summoned the lower orders to do the fighting, and dying, for them.

The robber-barons' main mouthpieces among the young men of Ulster – Carson, Crawford, Craig – came festooned in flags to the sound of the drum and whooped it up for king and country. Tens of thousands answered the call.

Meanwhile in Nationalist Ireland, John Redmond had the green flag wrapped around him and 'Ireland Boys, Hurrah!' on his lips as he pledged freedom at last in exchange for service to the Empire. Tens of thousands rallied there, too.

Together at the Somme, as at Messines, Ypres, Cambrai,

etc., they were flung to their death by the fistful, like chaff to a capricious wind.

They died for nothing worth a drop of the sweat of one working-class person.

Why is there such hushed remembrance of World War One? Because those who contrived it and their epigones, need the truth of it hushed.

Why do they shroud its dead in such determined sentimentality? The better to obscure the obscenity they wrought.

We should remember forever all who died at the Somme, and all across the killing ground that Empire made of Europe, and freshen our memory anew each year, lest we forget the evil of it all.

We should have in our mind's eye in the season of Remembrance not the clean-cut youth of the gable murals rushing to death for God and Ulster; we should see him instead, from the Shankill or the Falls, as he likely was.

And watch the white eyes writhing in his face,
His hanging face, like a devil's sick of sin;
If you could hear, at every jolt, the blood
Come gargling from the froth-corrupted lungs,
Obscene as cancer, bitter as the cud
Of vile, incurable sores on innocent tongues, –
My friend, you would not tell with such high zest
To children ardent for some desperate glory,
The old lie: *Dulce et decorum est*
Pro patria mori.

Gerry McDonnell

Gerry McDonnell was born in Dublin in 1950. His poems have appeared in *The Honest Ulsterman,* and *Salmon.* His poetry collections are: *From the Shelf of Unknowing and Other Poems* (1991) and *Mud Island Elegy* (2001). In July 2000 his stage play, *Making it Home*, was performed at the Crypt Arts Centre, Dublin Castle.

Collateral Damage

(front page, *Boston Globe*, 3 December 2001)

Hassad Saed sits
At the bedside
Of his young nephew,
Hands covering his face.

The boy, Noor Mohammed
Lies delimbed, in the dark,
White bandages wound tightly on the truth –
No hands and no sight
In this lifetime.

Let us not lay at God's feet
That He calculates the cost of war.
Let us pray that Mohammed
Has in his heart
What he has lost.

Cambridge, Mass., 7 December 2001

Oisín McGann

Oisín McGann has spent most of his working life as a freelancer, serving the advertising, design, animation and publishing industries in a range of roles. Brought up in Dublin and Drogheda, he spent several years in London, but returned to live in Ireland in 2002. His two novels, *The Gods and Their Machines* and *The Harvest Tide Project*, are due to be published soon by The O'Brien Press.

In the following extract, a teenager named Benyan Akhna hides in the false bottom of a small crate to smuggle himself into the city of his enemies. He joined the rebel group, the Hadram Rahni, as an angry boy and they have turned him into a well-trained and vengeful young man. Now, as he waits for the moment when he will share a supernatural death with his targets, his zeal is shaken by doubt and fear and by the ghosts who travel with him. The organisations, places and characters mentioned in the extract are fictional.

Left alone in the enclosed darkness of the false bottom of the cargo crate, Benyan lost track of time. It seemed an age since he had felt the box being lifted up from the concrete and carried bumping and swaying, to be set down on a different-sounding floor, the wooden boards of the train. Then the train had moved off and he could hear the rumble of its engine and the clack-clack of the steel wheels on their tracks. The birds in their baskets twitched and rustled above him, increasing his sense of isolation. And in the crushing, lonely darkness, the spirits made themselves felt.

All the history his father and uncle had taught him came to life in that tiny, black space. Visions of families burnt alive in their homes for not paying rent to absentee landlords, Shaneyans being branded as criminals and hung for practising their religion, and ancient memories of cannon smashing rebel armies armed with swords and knives. There were the markets where men, women and children were bought and sold like cattle for the slave trade and transported to other countries in slave ships, kept in brutal conditions where nearly half of them would die before they reached their destination. With the spirits threading through his consciousness, he relived these events as if he had been there.

At one point, a cramp in his calf brought him back to his own situation and he twisted and contorted to try and stretch the leg out to ease the pain. He managed to quell it slightly by forcing himself into one corner and pushing his foot into the opposite corner, massaging the muscle until the ache died away. In the minutes that followed, he reflected on what he was going to do in Victovia and a part of him wondered about the people he was to kill. Was the death of one person really a reason for causing the deaths of others? And if that was so, where did it stop? The toll of deaths in the history of this conflict was beyond measure; if each killing automatically demanded retaliation, how could it ever end? The books of Shanna allowed for revenge, but they also preached its futility. He closed his eyes to pray for guidance, but as if this opened the door, the spirits came back and plunged him once more into the crimes of the past. And with each vision, he felt a piece of himself being torn away and lost to the pain.

Benyan felt as if a horde of ghosts had passed through his head, but these new ones were different. He could make out individuals, there were personalities who were making themselves felt, eager to share their grief with him. The more time that passed, the less control he had over his body and mind and he mourned the loss. He closed his eyes wearily and succumbed once more to the force of the spirits' memories.

He found himself enmeshed within the minds of three men

and two women, brothers and sisters of the Lentton family, his consciousness sharing all five bodies at once. He could not distinguish between himself and each of the separate souls of this family. He did not question how it was possible; there was no need, it simply was. They lived in the town of Exurieth, which lay in the shadow of the plateau city of Mauraine. It was a Bartokhrian town in Altiman territory and it had been the scene of protests for years. The waste from factories in Mauraine was contaminating the farmland around Exurieth, poisoning its crops, killing livestock and spreading sickness through the town's water supply. In court, the lawyers acting for the factory owners had proved, in theory, that this could not be true. But the crops and the animals still died.

The Lenttons responded by going out one night and filling some of the pipes that spilled the waste with bags of cement. Benyan enjoyed taking part in the sabotage, dressed in fisherman's waders, waterproofs and heavy gloves to protect his skin from the toxic slurry. They worked quickly to stack up the heavy paper bags of cement while the flow had slacked off. The flow would start again slowly enough to soak into the cement and create a serious blockage. When he was done, he congratulated each other and slapped each other on the backs. They were proud of himself, his town was not going to let themselves be pushed around by the faceless men in suits. The contaminated waste in the blocked pipes backed up and four factories in Mauraine had to be evacuated the following day because of the poisonous fumes that filled the buildings.

A few nights later, the five of him got together again and set out to block the remaining pipes that were contaminating their land. One of him kept watch while the others unloaded the bags of cement. The fumes in the first pipe they came to were different from the others, they smelled of petrol and some kind of oil-based chemicals. His torches waved around the dark opening in the wall of a ditch, taking in the slow trickle of the chemical slurry from the gaping mouth of the corrugated pipe. All five of him covered their mouths with wet neckerchiefs to help against the fumes as they stacked the heavy sacks.

They were half way through when he was suddenly blinded by the beams of three spotlights. They were surrounded, men armed with guns stood on the walls of the ditch, hidden in the shadows behind the lights. He shielded his faces from the glare, trying to identify the silhouettes who shouted down to them. They ordered all of him to finish blocking the pipe. Frightened and confused, he did as they were told, shoving handfuls of loose cement into the gaps that remained between the bags. The five Lenttons turned to face the men behind the light, wanting to get out of that ditch and get back to his homes. Then someone lit a flare and he realised with sudden terror that they were standing knee-deep in a pool of the petrochemical slurry. The flare was tossed into the ditch and Benyan died five burning deaths.

He was screaming when the top of the box opened and three pairs of strong hands pulled him up into a light that scorched his eyes.

'Turn off some of the lights,' a voice said. 'By Shanna, this one's far gone.'

He fell to the floor and thrashed around, still suffering from the flames that had disappeared with the vision. Soon, he was able to feel the cool floor and his eyes adjusted to the dim light. But Benyan Akhna no longer looked out from those eyes. Having driven themselves between his mind and his senses, the Lenttons had taken residence. Five vengeful spirits looked up from the floor of the dark freight depot, hungry for others to share their fate.

Medbh McGuckian

Medbh McGuckian was born in Belfast in 1950. She has written several collections of poetry, including *The Flower Master* (1982), *Venus and the Rain* (1984), *On Ballycastle Beach* (1988), *Marconi's Cottage* (1991), *Captain Lavender* (1994), *Selected Poems* (1997) and *Shelmalier* (1998).

River of January

I do not sing of arms and the man,
I have nothing to say which I can say.
People walk about as if they own
Where they are and they do.

But those separated by a forest
Of error from the separated,
Call the deadly loneliness
By many other names.

How and why do we dream
Of living in unity
With the island that has become
As the forest was, warming our hands

In her warm breast? Creating
a flower-rich shelter or the heart
Of a nest, in a wood left untouched
By the prospering suggestion of orchards?

It is just a promise and strong
But beautiful like every promise
In its pure mountain form
Whose shadow was the forest cleared.

The old meanings of the forest
Read the seasons on the seasonless,
Treeless, herbless, overfished
Sea bespread with eye-driven ships,

And the fixed salts of plants
In the sand which paves the sea.
They fall into a light-bringing
Language, till all the light

Within the light, deceptive
As the leaf fall of the day
From a tree that has died in sleep,
Satisfies, by its quality of chaliceness.

Bernard MacLaverty

Bernard MacLaverty was born in Belfast in 1942. His short-story collections include: *Secrets* (1977), *A Time to Dance* (1982), *The Great Profundo* (1987) and *Walking the Dog and Other Stories* (1994). Novels include: *Lamb* (1980), *Cal* (1983), which were both adapted for screen, *Grace Notes* (1997) and *The Anatomy School* (2001). *Grace Notes*, from which this extract is taken, was published by Vintage Press, 1997.

from *Grace Notes*

Catherine McKenna is a composer from the North of Ireland and she is flying back to her home in Scotland after her father's funeral in mid-Ulster. She remembers winning a travelling scholarship to visit the composer, Anatoli Melnichuck, in the city of Kiev. Olga is his wife and translator.

They went back to the Melnichucks' house and Olga cooked breakfast. Afterwards Anatoli played Catherine a tape of a work he had recently recorded.

'A Hymn to the Mother of God Seated on the Throne of Heaven,' said Olga. The tape machine was ancient, totally unable to reproduce the sounds he had given to the choir. But still the magnificence came through. It was spare and structured and utterly memorable. When the last notes died away the tape hiss continued and the dog's tail thumped on the rug. Olga said, 'Since the political changes he can give his work the names he has inside his head. Ten years ago he would call it Chronos Four or something. Shostakovich said to him once, "Your head is a safe. Only you know the numbers to open it."'

'You knew Shostakovich?'

'Yes, Dmitri Dmitriyevich came here to Kiev with his wife number three, Irina Supinskaya. To discuss his Babi Yar symphony. It was very brave music to write at the time. You know Babi Yar?'

'No, not . . .'

'Maybe you are too young. Babi Yar is a place of death. In 1941 the Nazis made all the Jews of Kiev come together and they took them to Babi Yar – thirty-five thousand – men, women, children – and they shot them and put them down in a ravine to be buried. Evtushenko wrote a poem and Shostakovich put it in a symphony. But the anti-Semites said not all the dead are Jews. There is Russians and other prisoners.'

Hearing the words Babi Yar and Shostakovich, Anatoli became agitated. He spoke to Olga.

'He says they were all Jews who were killed. Dmitri Dmitriyevich was right – we must all fight anti-Semitism. The beginning of anti-Semitism is talk, is hatred – the end is Babi Yar. But, of course, in 1962 the State said there is no anti-Semitism in the USSR.'

The stewardess came and stacked the empty meal trays, crowded them one on top of the other and pushed them into her trolley. Catherine thought of the geography of the places of death in her own country – it was a map which would not exist if women made the decisions – Cornmarket, Claudy, Teebane Crossroads, Six Mile Water, the Bogside, Greysteel, the Shankill Road, Long Kesh, Dublin, Darkley, Enniskillen, Loughinisland, Armagh, Monaghan town. And Omagh. And of places of multiple deaths further to the east – Birmingham, Guildford, Warrington. It was like the Litany. Horse Guards Parade. Pray for us. Tower of London. Pray for us. Alone or with others. For the dead it didn't matter how many companions they had or where it happened. It was awful to think that if she wrote the most profound music in the history of the world it would have no effect on this litany – it would go on and on adding place names. Once in the University library looking through the *Encyclopedia Britannica* of 1911 for something on the sonata form – the eleventh edition had been recommended for its excellent philosophical and musical contributions – she came across Somme. It was described as a department of northern France – great rolling plains generally well cultivated and fertile etcetera etcetera. And that was all. Yet somehow she knew that her act of creation, whether it was making another person or a symphonic work, defined her as human, defined her as an individual. And defined all individuals as important.

Liam MacUistin

Liam MacUistin was born in Dublin in 1929. His plays include: *Post Mortem* (1971), *The Glory and the Dream* (1982). His work was selected for inscription in the National Garden of Remembrance, Dublin. His novels include: *Esperanza* (1995). His books for young readers include: *The Táin, Celtic Magic Tales, The Hunt for Diarmaid and Gráinne*, all published by The O'Brien Press.

Hope

Skibbereen,
February 14th, 1847

Dear Robert,

I have been a month here now surveying the effects of the famine on this area and reporting back to the Central Relief Committee of the Society of Friends. My friend and fellow-Quaker, Richard Hunt, is here with me. A sterling man but somewhat nervous in temperament. He has a deep fear of our being attacked by the starving people as we travel around in our coach. But it is impossible to avoid the starving wretches even if I wished to. Whenever we stop they come surging around our coach begging for food. I give them whatever help I can. I try to assure Richard that we are in no danger.

Despite their extreme poverty and hunger these people are peaceful and law-abiding.

I am afraid I gave Richard an additional cause for worry today. We were travelling in our coach on the road to Queenstown when we came upon this little waif sitting by the roadside. He did not seem to notice us as we approached him. His large blue eyes stared into space as though they were contemplating another world.

I ordered the driver to stop. Richard immediately urged me to continue on but something in those bright staring eyes made me stay. I got out of the coach and spoke to the boy. His eyes focused on me and he mumbled some words only one of which I understood: '*Ocras*', which I believe means hunger in the Gaelic tongue.

I lifted him into the coach, much to the disquiet of Richard who muttered dire warnings about fleas and fever and our lives being placed in peril.

I brought the boy back here and had him fed, washed and given new clothing. He is like a little angel with his blond hair and blue eyes. Richard wants me to place him in the local workhouse but I think he deserves a better refuge than that. If I cannot trace his family or relations (who, in all probability, are now dead) I may consider adopting him legally and giving him my own name. Since I have no children, and am unlikely ever to have any of my own, my adopting this boy would ensure the continuity of my name.

In the meantime, Richard is talking of returning to England! With sincere good wishes,

Gilbert Fry.

* * *

The boy sat at the side of the dusty road, his chin resting on his right fist. He had walked many miles that day. Now he could go no further. His stomach was empty and his throat was parched. He had tried chewing some grass to ease his hunger but swallowing it made him sick. The only water available was in a cesspool further up the road where a crowd of emaciated refugees had established a camp on the hard ground.

There had been no lack of water in the wells and streams near his village. But the village had been destroyed, burned to the ground by the mob of Hutus incited to hatred by political manipulators. They had killed his parents and brothers and sisters. He had managed to escape into the bush and joined a column of Tutsi refugees fleeing over the border. But, weakened by hunger and thirst, he had dropped behind and finally sagged to the ground.

He sat there now, his huge brown eyes staring hopelessly back along the road. He stared into emptiness, not noticing the Range Rover that came over the hill leaving a dusty slipstream in its wake. It sped past the boy, then suddenly braked to a stop. It reversed back to where he sat and the driver got out and came over to him.

Rwanda,
September 17th, 1994.

Dear Sis,

I am sorry for not writing to you sooner but I have been extremely busy with field work among the refugees since arriving here.

Regrettably, the war situation has not improved and I fear that thousands more of these unfortunate people will die of hunger and disease unless the flow of aid can be considerably increased. Next to food the main problem here is water. Most of what is available is polluted and dangerous to drink.

Now, on a more optimistic note: would you like to become an aunt? I am seriously thinking of adopting a young orphan boy I found starving by the roadside today. I was travelling on my way back to camp when I spotted him sitting there, his brown eyes staring forlornly into space. I brought him back to camp where I fed him and got one of the medics to check him out healthwise. Apart from malnutrition he appears to be all right.

If I do decide to adopt him I will probably give him my own name. Or is one Gilbert Fry enough to have in the family? On second thoughts you needn't answer that question.

Love,

Gilly.

Aodhan Madden

Aodhan Madden was born in Dublin in 1954. A playwright and scriptwriter, his works include: *The Midnight Door* (1983), *The Dosshouse Waltz* (1985), *Sensations* (1986), *Private Death of a Queen* (1986), *Remember Mauritania* (1987), *Obituaries* (1991), *Ladies in Waiting* (1993) and the screenplay for *Night Train* (1998); *Demons*, a collection of verse (1981), and *Mad Angels of Paxenau Street*, a collection of short stories (1990). He has also had numerous stories broadcast on RTÉ. 'Remembering the Somme' was first published in the *Irish Press*, 1986.

Remembering the Somme

On the seventieth anniversary of the Battle of the Somme I met a survivor of that carnage in July 1916. He was Jimmy O' Brien, one of the last of the Dublin Fusiliers. He sat in his optician's shop in Aungier Street and quietly reflected on a war that haunted his entire adult life. The Somme was one of the bloodiest battles in human history. Tanks were used for the first time and tens of thousands of teenage boys and young men were slaughtered in one single day.

'It started about 6.30am,' said Jimmy. 'Suddenly seven thousand guns opened fire and the whole earth shook. We were stunned most of the time. But the thing I remember most was that colossal noise. It was awful ...'

What I remember most about my interview with Jimmy O' Brien was the powerful resonance of his silence. It was as if the very name, the Somme, was explanation enough. All the horror of mechanised warfare is conveyed by this name.

We think of death and suffering on a scale unimaginable in prior human history. We see hundreds of thousands of shattered human bodies strewn across ditches and barbed wire. We hear the gigantic war machines make their debut on the world stage, the wildest imaginings of HG Wells come to pass in a few bloody fields of Belgium. There are other names too amid the debris of the twentieth century – Auschwitz, Nagasaki, My Lai – which reduce us to horrified numbness ... But the Somme set the standard, as it were, for a century that would see man sink into unspeakable barbarism just as he was reaching out to touch the moon.

When I asked Jimmy about the conditions in the trenches, he was slow to respond. It was difficult for him, even seventy years later, to talk about the horrors he had seen. 'We couldn't kill the rats,' he said, 'because they would get diseased. So they grew as big as cats and they ate the dead and the nearly dead soldiers ...'

Jimmy fought alongside the legendary Fr Willy Doyle from Gardiner Street, who was chaplain to the Dublin Fusiliers. 'He was blown to bits right in front of me in 1917.' Fr Doyle was a hero. He took great risks to save his men, often exposing himself to German fire in order to drag a wounded comrade back into the trenches. He was awarded the Victoria Cross posthumously and now lies in Flanders Field alongside thousands of his countrymen. When I asked Jimmy if he ever returned to Flanders to visit the graves of the war dead, he shook his head. 'I saw too much when I was there,' he said. He showed me some old, yellowed photographs of his friends who were killed in the Somme. They look like schoolboys. They had that haunted look that Francis Ledwidge and Rupert Brooke so vividly described in their poems about the war. In fact, the little office at the back of his shop was a shrine to his lost companions. On one wall hung pictures, flags, medallions – old, frayed memorabilia of the Fusiliers and of the war. We sat for ages in that sacred place listening to the Dublin traffic groan and snarl outside. Occasionally Jimmy would recall an incident or an anecdote like the one about Queen Victoria's remark about the Dublin Fusiliers: 'Now them is men,' she is supposed to have said! But mostly it was still too painful for Jimmy to dwell on the awful detail of the Somme, or Ypres or Mons or Passchendaele.

'What can I say,' he said a few times. There were no words left to describe the sheer scale of the slaughter and the obscene folly of sending millions of young men to fight tanks and machine guns.

So he was left with the simple dignity of silence. Jimmy died some time after our meeting. He was not to see Official Ireland recognise at last the Irish war dead of Flanders. But he kept his own watch over the memory of his lost friends and this seemed to me to be deeply moving.

Paula Meehan

Paula Meehan was born in Dublin where she still lives. She has published five collections of poetry, the most recent being *Dharmakaya* (2000), from which the poem published here is taken. 'Fist' appears by kind permission of Carcanet Press, Manchester.

Fist

If this poem, like most that I write,
is a way of going back into a past
I cannot live with and by transforming that past
change the future of it, the now
of my day at the window watching
the coming and goings to Merrion Square,
then, when you present your hand to me
as fist, as threat, as weapon,
the journey back to find the hand of the little child,
the cupping of her balled fist
in my own two adult hands,
the grip of her fury, the pulse at her wrist
under the thin thin skin,
the prising loose of each hot finger
like the slow enumeration of the points of death
and the exact spot that I will have kissed
where the fate line meets the heart line –
my bloody mouth a rose suddenly blooming,
that journey takes all my strength
and hope, just as this poem does
which I present to you now.

Look! It's spread wide open in a precise
gesture of giving, of welcome,
its fate clear and empty, like the sky,
like the blue blue sky, above the Square.

Sinead Morrissey

Sinead Morrissey was born in Northern Ireland in 1972. She is currently
Writer-in-Residence at Queen's University, Belfast. She has published
two collections of poetry, *There Was Fire in Vancouver* (1996) and
Between Here and There (2002). 'The Wound-Man' was commissioned
by the Royal Festival Hall for their Poetry International Festival 2002.

The Wound-Man
for Federico García Lorca

It would have been a kind of action replay,
only worse. The white handkerchiefs.
The unimaginable collapse. The day
the markets crashed and unleashed
unknowing through the New York streets

saw you transfixed, a witness in Times Square,
as the world went down in hysterical laughter
and diminishing shrieks. Then thudded over.
All hope in the gutter, blooded and lost. How you loathed
the reflections of clouds in the skyscrapers

and the glittering rings of the suicides.
It was all one in New York: the manacled roses, oil on the
     Hudson,
financial devastation. Had you survived,
Federico, say, Franco's henchmen,
or the war that was to open like a demon from his person,

or the later war, and all the intervening years
between that fall of faith and this, what would you think?
Would you know what has happened here,
the way we do not know what has happened? Where
would your fury go? We shiver on the brink

of an ending, and a war stretches in front of us,
we stand where you stood. As for me,
I see the Wound-Man walking, tall and imperious,
through the streets of America, surly
and muscular, from the textbook of Paracelsus.

He's been badly hit. There are weapons through every part of
     him.
A knife in the cheek; an arrow in the thigh;
someone has severed his wrist bones, on a whim,
and thrust a sword into his eye.
They've flung razors at his flesh to pass the time.

And yet he rears. Sturdy and impossible. Strong.
Loose in the world. And out of proportion.

Danny Morrison

Danny Morrison is a former national director of publicity for Sinn Féin. After his release from prison in 1995 he became a full-time writer and commentator. He is the author of three novels, a collection of letters and a memoir. He lives in Belfast. 'Nayirah' was published in *Andersonstown News*, 10 February 2003.

Nayirah

She gave her name as Nayirah and it was explained that she couldn't be fully identified as her family was still trapped in Kuwait. Nayirah was fifteen years old and was speaking before a Congressional Human Rights Caucus, just months after Iraq had invaded Kuwait in August 1990. She said she was a refugee who had been working as a volunteer in a Kuwaiti hospital throughout the first few weeks of the Iraqi occupation.

'I volunteered at the al-Addan hospital. While I was there I saw the Iraqi soldiers come into the hospital with guns, and go into the room where fifteen babies were in incubators. They took the babies out of the incubators,' she sobbed, 'took the incubators and left the babies on the cold floor to die.'

This horrific story became the lead item in newspapers, on radio and television, not just in the USA but across the world. Amnesty International took out a full-page advertisement condemning Iraq. Six members of the US Senate said that this was good enough reason to go to war. President George Bush the First cited the incident in his speeches. It was repeated at the United Nations Security Council when Dr Ebrahim, a surgeon, stated that he had buried forty babies pulled from the incubators by the Iraqis.

On 29 November 1990, the UN authorised use of 'all means necessary' to eject Iraq from Kuwait. By 12 January 1991, when the US Congress authorised the use of force against Iraq, the number of premature babies removed and left on the cold floor to die had climbed to 312.

Nayirah, in fact, was a member of the Kuwaiti royal family and the daughter of Saud Nasir al-Sabah, the Kuwaiti ambassador to the United States. She hadn't worked in a Kuwaiti hospital and was in the USA when the Iraqis invaded. And Dr Ebrahim was a dentist, not a surgeon, who, after the war, when the scam was exposed, and 100,000 Iraqis had been killed, admitted that he had never buried any babies or seen any.

Their stories were fabrications. Nayirah had been coached in her testimony by Lauri Fitz-Pegado of the Public Relations company, Hill & Knowlton, hired by the Kuwaiti government to win American support for the war under the rubric 'Citizens for a Free Kuwait'. It didn't matter that the Emir of Kuwait, whom the USA was going to reinstate, had, four years earlier, disbanded the token national assembly and censored the media. Free Kuwait sounded good.

Lauri Fitz-Pegado was a former Foreign Service Officer at the US Information Agency. She had previously been a lobbyist for the dictator Jean Claude ('Baby Doc') Duvalier, President of Haiti.

Hill & Knowlton's president, Craig Fuller, was one of George Bush the First's closest friends and political advisors when Bush was vice-president under Reagan.

In September 2002, US television ABC's 'Primetime Thursday', interviewed Parisoula Lampsos, a fifty-four-year-old woman of Greek extraction who had lived in Baghdad most of her life. She is one of the sources used by Pentagon intelligence officials who claim that Saddam Hussein has links to Al Qaeda, links which are being used by Bush and Blair to justify the forthcoming war.

Parisoula claims that she was Saddam's mistress, that he was a Viagra enthusiast who enjoyed listening to Frank Sinatra singing 'Strangers in the Night', as well as torture victims crying for mercy. She says she watched him preen in front of a mirror declaring, 'I am Saddam. *Heil* Hitler!' She said that she had once seen Osama Bin Laden at Saddam's palace and that in the mid 1990s Saddam had given money to him.

'The first casualty when war comes, is truth,' said American Senator Hiram Johnson in 1917. When Britain was fighting the Boer War the British press carried hundreds of atrocity stories, including one about Boers attacking Red Cross tents while brave British doctors and nurses were treating the wounded. Documentary footage caused outrage when shown in British cinemas. But it was completely false and was shot on Hampstead Heath in London, using actors.

In September 2002 Tony Blair published a dossier on Iraq's chemical and biological programme, yet the CIA believes that the source of that report – an Iraqi defector – cannot be trusted and might be 'embroidering' his story.

Recently the BBC Radio 4's 'Today' programme quoted sources from British Intelligence taking issue with the way Tony Blair was distorting their reports in order to make a case for war. Sections of the CIA have made similar allegations that its reports are being 'cooked' by the Pentagon to sustain a case which may not exist.

On 5 February 2003, the US Secretary of State, Colin Powell, made his presentation to the UN about Iraq's alleged chemical, biological and nuclear weapons' programme and links to Al

Qaeda. In the course of his speech he called attention to 'the fine paper that the United Kingdom distributed … which describes in exquisite detail Iraqi deception activities.' But within twenty-four hours it emerged that large parts of the British 'intelligence' dossier were a 'cut-and-paste' job, taken from published academic articles, some of them ten years old, indicating a real dearth of fresh and original intelligence conclusively showing that Iraq actually possesses weapons of mass destruction.

I watched Colin Powell interpret satellite photographs. He claimed they showed Iraqis cleaning up a chemical munitions bunker and their lorries taking away the contraband. Why didn't the satellites track the lorries and reveal their location to the UN weapons inspectors? If Saddam Hussein has chemical and biological weapons, why didn't he use them against US and British forces when they drove him out of Kuwait? Why didn't he pass them on to Al Qaeda during the past twelve years of humiliation, when the no-fly zone was imposed on his air force, and when half a million Iraqi civilians were dying because of the embargo and through the spread of cancerous diseases from the depleted uranium shells fired by the US and Britain during the Gulf War?

We have been lied to repeatedly so we cannot now believe George Bush or Tony Blair.

Jean O'Brien

Jean O'Brien's work is widely published in magazines and journals and has been broadcast a number of times on RTÉ radio. She has two collections of poems: *Working the Flow* (Lapwing, 1992) and *The Shadow Keeper* (Salmon, 1997). A founding member of Dublin Writers Workshop, she is currently working on a new collection of poems entitled *Dangerous Dresses*.

War

The wasps have been murmuring all summer,
now they are garrisoned in the attic
for a last-gasp attack. Daily we find bodies
dead or wounded on the bare boards
or stairs. We wear slippers now for fear
of stings. We wonder do they remember
the sun's dazzle as they buzz towards
electric sun lights, then stunned, fall back
retreating into oncoming winter.

When we think they have finished their forays
we will climb into the attic, pull our
selves over the ceiling's threshold
into the cold air under the grey slates
and rafters that make up their sky.
We will mop up any stragglers,
seek out the nest, slash and destroy,
head them off from a reconnaissance
this time next year.

Mary O'Donnell

Mary O'Donnell was born in County Monaghan. A novelist and poet, her works include *The Light-Makers* (1992), *Spiderwoman's Third Avenue Rhapsody* (1993) and *Unlegendary Heroes* (1998). *The Elysium Testament* was published by Trident Press, 1999.

from The Elysium Testament

John spoke recently about survivors of the Holocaust, the ones who witnessed the worst. Not all have been able to bear the rhythms of seasons, the passing of time, the waxing and waning of moon, the passage of the sun from the Tropic of Cancer to the Tropic of Capricorn and back again, or even the cycles of their own bodies. Not all were able to pick up the shards of a broken peace.

We went to the zoo but it began to rain, so we sat in the tea-rooms, surrounded by a party of boys who saw us as a species to gawk at. Around us, the windows steamed up. In the end, I stuck out my tongue at one of the boys who promptly told me to *fuck-off Missus.*

John spoke softly. On that day, he had shed his harassed look. He was relaxed and intent on enjoying an afternoon off. He had the knack of making everything he said sound intimate, and made me feel as if I should be anyplace else but a zoo tea-rooms, surrounded by apes of boys.

'It isn't always possible,' he remarked, pressing his fingers to my palm, 'to say to yourself, from now on, I'll look at the world and relearn the rules.'

'Which ones?' I asked, unimpressed.

'The ones to do with contentment. Maybe endurance too.'

The list of those who could not endure the deceptions of a would-be happy world is long. Primo Levi flung himself to his death from his apartment block, years after the healing work of writing, long after *The Drowned and the Saved*, when the work of intellect and reason was not strong enough to conquer the torments of memory.

We left the zoo and walked beneath the dripping trees of Phoenix Park. We talked about murderers and rapists, rent-boys and prostitutes and politicians, the way things happened in that huge park-land, how you had to choose your time for walking and jogging. In a way, I was as happy as I'd been in a long time. It was as if I'd fallen into a little light bubble. I think he was happy too.

The deaths of children survive like scars, whether it is the deaths of the 4051 boys and girls who became known as the Children of Drancy in the summer of 1942, or the personalised, intimate deaths of children known to us, which never make it into the history books, and are irrelevant to sociology, history or statistics. The Children of Drancy were kept for four days without food at the Velodrome d'Hiver on the outskirts of Paris, their mothers were removed, then they were loaded three hundred at a time into cattle trains at the Gare d'Austerlitz and taken to Auschwitz.

Ciaran and I were at the Gare d'Austerlitz. That channel of horror is used by busy, contemporary people, people in love, people half-mad, people rich, poor, ignorant, funny. If some of the French-Jewish parents of those children survived, what I want to know is, for how long? For how long can adults stand the knowledge of children's pain?

The Children of Drancy read books like *Babar*, they played Housie with one another, they had pretend adult parties.

Roland liked six teddy-bears in his bed. He collected unripe plums the colour of greengages from the tree at the end of the garden and stored them until they shriveled in the back of his tricycle. He liked chips, sausages and Coke, apricots, Jellytots and raisins, the illustrations in a book of the Edward Lear poem 'The Owl and the Pussycat', and watching Chinese ping-pong competitors on EuroSport.

I can't compare the fate of the Children of Drancy to our child's death. All I hope is that he knew nothing, that he was completely unconscious. But he is gone, and that knowledge is as raw to me as it must have been to the parents of those children. Their presence is scarcely remembered in the annals of history. His will never be forgotten in the slim passage of our lives.

I no longer want to have to remember, but cannot forget.

Sheila O'Hagan

Sheila O'Hagan is a poet and short-story writer. She has published two collections of poetry with Salmon Publishing: *The Peacock's Eye* (1991) and *The Troubled House* (1995). Her awards for poetry include the Patrick Kavanagh Award (1991), The Tribune Hennessy Award (1993) and the Strokestown International Award (2000). She is currently guest editor of *The Cork Literary Review*. 'Home from the War' was published in *The Troubled House* by Salmon Press, 1995.

Home from the War

Something was always happening in Crewe
When my father came home on *The Irish Mail*.
Sometimes they missed *The Princess Maud*
And it took him twice as long. Could be

He needed to rest awhile, forget
The jeep rides over deserts, seas of Naffi tea,
A common hatred of Montgomery.

Once he sent me a photo of George Formby
And his little dog – both entertained the troops.
My father also played the ukulele. It went well
With his cheery whistling, his matey jokes:
'If the sky fell we'd catch larks.' They sent
His money home, what little he was paid
For fighting Hitler and the Iti Brigade.

'Reported missing, presumed dead,'
Best joke of the year for him, home on leave
And sipping tea when the telegram came.
'That's the army for you,' he said,
But for all I saw of him he could have been.
Twice he was nearly killed. My mother
In a new dress, had had her hair done.

Driving a truck, he got the speed wobbles.
I couldn't understand how in all that sand
It mattered. And a Red Cross plane dropped
Like a stone before the second engine,
At eight hundred feet, took them out of it,
He strapped to a stretcher with sciatica
And the 'bronichals', all from the heat.

In Rome the opera still played. He cried
At *La Bohème*. And cried to see the children
Without food. Gave them his chocolate ration.

At home my mother cried because the PO
Lost her paybook. Said she couldn't cope.
In nineteen forty-five he came home,
Went fishing in the river, had a heart attack.

It could have been the Players' cigarettes,
Could have been the canteen gin although
He pleaded war strain. Six medals from the King,
A pension of seventeen shillings a week
For seven years away from family and home.
My mother went to work to raise a headstone.

Siobhán Parkinson

Siobhán Parkinson has published several books for children and young readers including *Sisters – No Way!*, winner of the overall Bisto Book of the Year award, *Four Kids, Three Cats, Two Cows and One Witch – Maybe*, Bisto merit award winner; her most recent book is *The Love Bean*; all published by The O'Brien Press. She edits *Inis*, journal of Children's Books Ireland. This extract from *No Peace for Amelia*, appears by kind permission of the author and The O'Brien Press.

from *No Peace for Amelia*

This is a children's novel set in a Quaker home in Dublin in 1916. Frederick, who is Amelia's beau, has signed up and is fighting in the Great War; Patrick, who is Mary Ann's brother, has been involved in the Rising and at this point in the story he is on his way to deliver an important message, but has been wounded and has ended up being looked after by the two girls.

'I have to go, I tell you, I must,' [said Mary Ann, meaning she will have to deliver the message on Patrick's behalf]. 'It would be worse for him to fail in this than anything. He'd rather die in the attempt and die with honour. Oh, Amelia, think if it were Frederick!'

Frederick hadn't been far from Amelia's thoughts. As she had bathed and bandaged Patrick, she had wondered if some girl somewhere might do the same for Frederick if she found him wounded in France. She imagined Frederick, sickly and shot, in a sweet-smelling haybarn on a French farm, and some apple-cheeked French farmer's daughter with a blue check kerchief round her head and strong peasant hands tending to her hero's wounds. She imagined it all so vividly that she was almost jealous of the French farm girl. But she didn't like it when Mary Ann mentioned his name, as if she had tuned into her private thoughts.

'At least Frederick is fighting in a proper, honourable war, not just a skirmish in a post office.'

'Oh!' cried Mary Ann, shocked into bitterness by Amelia's words. 'Honourable! Is that what you call it? What's so honourable about crouching miserably in a muddy, lousy trench and taking potshots at other miserable, muddy, lousy soldiers, and all for what? To keep England powerful, that's what for.'

'It's not!' said Amelia passionately. 'It's to defend Europe against the Germans. It's to safeguard the women and children of Belgium and France. That's what it's for, and it is honourable, it is!'

She stamped her foot, as she used to do when she was a child and was overcome with rage and indignation. For a moment, neither girl spoke. They faced each other over Patrick's prostrate body, Amelia white with anger, and Mary Ann's face dark and glowering. At this point, Patrick opened his eyes and looked enquiringly from one to the other, but neither of them noticed.

Silence crackled in the air between them. Minutes passed, and Patrick woke up properly. He lay and watched the two girls, trying to piece together where he was, what was happening. He felt for his gun, and then remembered that he had ditched it.

'I'm sorry, Amelia,' said Mary Ann at last, looking her friend in the eye. 'I'm sure whatever about the ould war, Frederick is honourable anyway.'

Amelia said nothing for another long moment. Then she relented and mumbled: 'I'm sorry too, I suppose. I ... I shouldn't have called your precious Rising a skirmish in a post office.'

'Well, I suppose you could call it that. But that doesn't mean the men and women too who are fighting in it aren't every bit as honourable as your Frederick. They're willing to sacrifice their lives for their country, and you can't do better than that.'

Amelia's mother would have replied that you could do better – that you could live for your country instead, and strive to make it a better place, but Amelia didn't say it. Instead, she just nodded and said: 'Well, I'm sure they are all honourable men, whichever war they are in.'

Then a thought occurred to her. It's war itself that is dishonourable. She turned this thought over in her mind. It was the first time she had been able to do what came naturally to Mama – make a clear distinction for herself between the war and Frederick's part in it. Yes, it's war itself that is dishonourable, she thought again ...

[Later, Patrick, in conversation with Amelia, says:] 'I suppose you're an Orangewoman. All Protestants are Orangemen.'

'No. I'm a Quaker.'

'Isn't that a sort of Protestant?'

'Yes and no. It's very different really. And we are neither nationalists nor unionists. We are pacifists.'

Patrick gave her a long, considering look, from his slate-grey eyes. She smiled at him, and then looked away in confusion.

'Anti-war?'

'Yes.'

'Ah, sure, aren't we all anti-war at heart! I mean, none of us *likes* fighting and killing.'

'It's not enough to be anti-war at heart,' said Amelia virtuously.

'What does that mean, now?' asked Patrick in a rather patronising tone that Amelia didn't like.

'It means,' she said firmly, 'that you have to work for peace, not just have a distaste for war.' Amelia surprised herself. She hadn't given much thought to what it meant to be a pacifist recently. She felt somehow that it might be disloyal to Frederick. But the arrival of Patrick, ill and wounded, in her own backyard, quite literally, had given her a new and less glamorous perspective on war.

Tom Paulin

Tom Paulin is a poet, writer, literary critic and lecturer in English at Hertford College, Oxford. His most recent book is *The Invasion Handbook* (2002). 'During The Countdown' was published in the *London Review of Books*, 20 February 2003.

During the Countdown

On the second day of the second month
2003
we were walking through Beeston
– it looked that Sunday
more like a wet northern

than a wet midland town
with big strange pollarded trees
on both sides of its not wide not grand
Imperial Road
– every single limbless hacked cutback trunk
was taller than the Victorian houses
and each a kind of *écorché*
displaced almost tarry with a blind scorched
halfconscious look
– these overgrown but somehow ambushed trees
they'd got too grand for a mere road
– maybe when their trunks were just saplings
it looked like an avenue in the making?
now these rooted
not quite cadavers were nearly speaking back
like a tamarack a hackmatack
– that is the American the charred larch

Gabriel Rosenstock

Gabriel Rosenstock was born in Limerick in 1949. A former Chairman of Poetry Ireland, his published collections of poetry include: *Self Portrait of the Artist as an Abominable Snowman* (1989), *Rogha Rosenstock* (1994), *Cold Moon: The Erotic Haiku of Gabriel Rosenstock* (1993) and *Syjójó* (2001). His tenth volume of poems, *Eachtraí Krishnamurphy*, will soon be published. 'An Maide Buataise' is a translation from German from the original by Michael Augustin.

Michael Augustin, poet and broadcaster, was born in Lübeck (Germany) in 1953. He has lived in Kiel and Dublin, and currently resides in Bremen. He has written for *Cyphers* and read at *Dublin Writers Festival*. His books have been translated into many languages.

Hans Christian Oeser has translated 'The Boot Tree' from Augustin's German. Oeser is a translator based in Dublin. He has translated works by many Irish writers including Brendan Behan, John F. Deane, Dermot Healy, Jennifer Johnston, Paul Muldoon and Patrick McCabe.

An Maide Buataise

Agus an maide a shíneadh an bhuatais?
Cad d'imigh air?
Uch, tá sé sínte in Auerstedt,
i bpáirc an áir, sin an áit ina bhfuil sé,
in Auerstedt.

Agus an chos a bhí sáite sa bhuatais
a shíntí ag an maide,
cá bhfuil an chos sin?
Tá sí sáite sa bhuatais i gcónaí
an bhuatais atá sínte sa pháirc in Auerstedt,
sin an áit ina bhfuil sí mar chos.

Agus cá bhfuil an duine
a bhfuil a chos sáite i gcónaí sa bhuatais
a shíntí ag an maide
is atá anois ina luí sa pháirc
in Auerstedt?

Níl sé i bhfad as seo -
i ngarraí na dtornapaí, go deimhin,
sínte faoin gcré

(Agus níl ach leathbhuatais air).

Hans Christian Oeser

The Boot-Tree

And the boot that was stretched by the boot-tree?
What became of it?
Alas, it lies near Auerstedt
in a field, that's where it lies,
near Auerstedt.

And the leg that was stuck in the boot
that was stretched by the boot-tree,
where has that leg got to?
It got stuck in the boot
that lies near Auerstedt in a field,
that's where it got stuck, that leg.

And where has that man got to,
whose leg got stuck in the boot
that was stretched by the boot-tree
and that now lies in a field
near Auerstedt?

He got stuck in a plot of turnips,
right next to it.
That's where he got stuck, that man!

(And has only one boot on.)

Ronan Sheehan

Ronan Sheehan was born and lives in Dublin. His works include *The Tennis Players* (1977) and *Boy with an Injured Eye* (1983). 'The Sack' was published in *Boy with an Injured Eye* by Brandon Press (1983).

The Sack

During the civil war which followed the defeat of the Germans, in a small mountain town in the Peloponnesus, an old man rolled a barrel full of human eyes along the street.

My father has blue eyes. My mother's eyes are brown.

Tall mountains line our valley on every side. The sun must climb high into the sky before he can spy on us. I feel my face grow warm and then I look upwards. No one ever comes or goes. They say there is a way, a very hard way. I have never seen it. I am walking along the path with my mother and father. It is an old path and there are big old trees on either side. My mother and father do not speak. Something has happened. They are worried. Everyone must go to the town.

We fall in with other people. Everyone hurries. My friend is walking with the old man. The old man is blind and must take care. The old man's cat is in a tree. My friend tries to make the cat come down. The old man likes to carry the cat on his shoulder. My friend wants me to help him. My father pulls me away.

My father walks beside his friend. The man's wife is close to my mother. The men say nothing and the women whisper. I look at the faces of the people in the street. They have strong faces. The wind and rain and heat make their faces hard. I have seen their faces laughing and talking, showing their teeth which are very white. Now their mouths are closed and their faces are harder.

The crowd gets bigger and we walk more slowly. We are going to the usual place, the big house in the centre. Many people are already at the door. A man stands on the step. There is a bundle at his feet. We find a place where we can see. The man is waiting for all the people to come. My mother stands beside my father. My father makes me stand beside my mother. The people press around us. I smell their sweat and I breathe the air they have used. I turn and see the old man standing at the back of the crowd. The cat is on the roof of a house. My friend is a little way from the old man. He doesn't know whether he should get the cat or stand with the old man. He waves to me. A shove in the crowd separates me from my mother. My father catches my arm and pulls me back.

The man on the step stands behind the bundle, then faces us. We are still. The man speaks but his voice is low and we can't hear.

He moves from the bundle and pulls away a blanket. It is a body. The arms and legs are tied. The head is limp on the shoulder. There is no hair and there are no eyes. The sockets are ringed with blood. The cheeks are stained with blood dried black.

There is a movement from the back. Someone shouts. My father shouts and so do I. People are screaming and pushing. Men rush from the crowd and attack the man by the bundle. They knock him down and kick him. A bone cracks and he screams. A man takes a stick from his pocket and jabs the eyes of the screaming man. Other men rush from the crowd and fight the attackers. My mother falls. The woman who walked with her screams and tries to run. My father hits her when she steps on my mother. He drags my mother off the ground. He catches me by the arm and pulls me. I try to run. A man punches me. I am falling but my father pulls me again and I run. We run over people who are lying on the ground.

We run along the street, bumping into people who are running against us. A man tries to stop us. He kicks my mother and he swings a big stick to hit me. He is a bigger man than my father. My father shouts at us to run faster. I dodge the stick and run after my father and mother. The man swings his stick at people running behind us. There are six or seven men at the end of the street. They have long heavy sticks. My father moves to the side of the street. There is a small house with the door open. A man lies in the doorway and a woman is trying to pull him inside. My father runs past her into the house. My mother follows and then I go in. We can't shut the door because of the man lying there. My father picks him up and drags him into the house. My father's arms are bloody. The man's arm is hanging loose. The woman holds the arm and the man makes a small screaming noise. The woman drops the arm and the man moans. My father and the woman lay the man on a bed. My mother helps me to shut the door. The man's hat is on the ground outside. My mother bolts the door and turns the key. I pull over a table. My mother sits on the floor and cries. The woman sits beside her. The man on the bed is still. His face is growing white. I wonder will his face be as white as my father's teeth. My father drags the bed across the floor. We put the foot of the bed against the table.

We all sit on the floor against the wall. There is no sound in the house but we can hear noise in the street. Footsteps are thudding on the ground. There are shouts and screams and the sound of breaking glass. Men are cursing and women are crying. I can hardly think or move. My face is against the woman's side. She turns and covers my head with her arm and presses my face in to her breast. I grip her leg with my hands. I can hear the blood pumping in her heart. Her leg is still as if it were dead.

The street grows quiet. My father crawls across the floor. We watch him. My mother stirs and whispers. He turns with his finger on his lips. There is a cupboard in the corner. He opens it and searches. He finds a loaf and a jug of milk. He crawls back to us, pushing the loaf and jug ahead of him on the floor with either hand. He drinks from the jug and takes a mouthful of bread. He tears the bread and gives a piece to my mother and a piece to me. He hands the rest to the woman. We try to eat easily but it sounds as if the walls are crunching. I feel better when I'm finished. The woman lays my head on her breast again. It moves slowly up and down. She strokes my hair softly and I fall asleep.

The window shakes. My father jumps up. He pulls my mother from the floor and shouts at me. They are running to the back door. I run. He is unbolting the door. I turn to see what the woman will do. She is rising from the floor. The light is poor but I can see that her face is nearly as white as the face of the man on the bed. She goes to the bed and lies on him. Two men are climbing in the window. They are big men with sticks.

My father and mother are outside the door. My father pulls me out. He gives me a key. They run. I have to lock the door. One of the men rushes at the door. I slam it shut and lock it. I want to run but I wait to see what will happen to the woman. I look through the keyhole. They drag her off the bed. She doesn't scream or fight. One of the men hits her hard on the head with his stick. She falls. The other hits the man on the bed. They tilt the bed to one side and the man rolls off. They lift her on to the bed. One of the men takes a knife from his pocket and cuts her dress in front from the waist to the ankles. Her legs are as white as her face, as white as my father's teeth. He takes off his coat and throws it on her head. The other lies on her, covering her legs. I can't see her.

I run down the lane. My father and mother are waiting. My father pulls me into a doorway. There is a gang of men with sticks down the street. They are banging a door. They break it and go in. We run across the street to another lane. We run down lanes and across streets until we reach the big tree at the edge of the town. We stand under the tree. My mother is crying. My father is trying to make her quiet. A voice whispers from above. I don't want to look up. The voice whispers my name. I look up into the branches. My friend and the old man and the cat are there. The old man has his sack with him. My friend waves to me to come up. I show them to my father. He pulls me and my mother away from the tree. We run as fast as we can along the path, the same one that we walked on this morning.

I am afraid of the trees. They are bigger and older than the people in the town. They know better who we are and where we are going. It is they that make big sticks for the men. Through gaps in the branches they let the moon and stars spy on us as we run.

I can hear noises around us. There might be men in the trees. It is pitch black in this part and I can't see. I trip on stones and sticks. I can hear my mother breathing in front of me. She must be close to my father. He knows the way.

They turn down the narrow path to our house. From the bend I can see home. The moon shines on the roof, making the wood seem like silver. Anyone who comes this way will see it and go to it. A tree would be safer. People are afraid to look into trees at night. You can see everything from a tree.

We lock the front door and shut tight the front windows. My father and mother lie on their bed in the back room and I lie on my bed in the front room. I wish the woman were with me. I could put my head on her breast and she would cover my head with her arm and I would fall asleep.

I go to the door of the back room. It is open. I can see inside. They did not cover the window. The moon is at the window, looking in. The room is silver. My father is lying on my mother. My mother's legs are as white as the woman's. My father's teeth are as white as my mother's legs. They can't see. They can't see me standing by the door in the dark.

I am lying on my dark bed. I am at the top of a tall tree in town. It is black tonight but I can see everything. Someone is walking along the street. It is a woman. Her hair is red with blood. Her dress is torn and trails behind her. Her legs are white and bleeding. She is going to the big house. She kneels before the man without eyes and hair. She unties the knots at his wrists and ankles. She raises him to his feet. She pulls him away from the step. She puts her hands on his bald head and stares into his empty sockets. She starts to sing. Softly first, then more strongly until the street is full of the sound. She dances and makes the man with no hair and no eyes dance. She sings and sings and they dance along the street.

I uncover the window. Everything is the same. The trees are many, and leafy, and tall. The mountains are behind the trees. The light is the thin grey light of dawn. It will be a long time before the sun climbs above the mountains. It will be a long time before I feel my face grow warm.

I can hear noises around the house. The front door is open. My father is sitting on the step. There is blood on his throat. His face is white. His eyes are gone.

The door of the back room is open. I can see my mother's legs. They are dead white. I do not go to the bed because I know that I will not be able to see her.

I am out in the grey light, walking. The trees do not frighten me now. The cat is creeping along the path in front of me. I follow the cat. The cat moves surely between the trees, over the sticks and stones. We come to the tree at the edge of the town. The cat climbs up. I crane my neck and see my friend and the old man. I ask them to let me come up to them. They whisper and then my friend waves me on. I climb up.

I am sitting on a high branch between the old man and my friend. The cat is on my shoulder. The old man opens his sack to show me what he has. The sack is full of eyes.

Peter Sheridan

Peter Sheridan is a playwright and theatre director born and living in Dublin. His works include *The Liberty Suit* (1978) and *Emigrants* (1978). 'The Glass Eye' is an extract from the play *Shades of the Jelly Woman* (1986), by Peter Sheridan in collaboration with Jean Costello.

The Glass Eye

I was eight years old when I went to sleep in me Granny's. Queen Street, number eight. Where the curtains were always pulled over to block out the light. Granddad couldn't stand the light. Since coming back from the war. The only light was a bare bulb covered with a brown paper bag. And the big brass bed was kept out from the wall. Because the bugs used to break through. Pop. To me it was like a fairy story sleeping there. I thought at first I was sent there because my own house was gone too small. Or maybe I was a growing girl among a lot of boys. But it wasn't any of that. I went to sleep there as a protection for my Granny, Nellser, against the glass eye.

SHE TURNS AND IMAGINES THE GRANDFATHER LOOKING AT HER.

Jesus, Mary and Joseph, stop staring at me. Granny, tell Granddad to stop staring. Tell him to lie the other way, on his side, tell him to lie on his side away from me. Jesus, there's stuff coming out of his eye, there, and there's stuff coming out there as well. The eye must be like a stone rubbing against the lid. And there's water coming out of his mouth in floods. Granny. And his lips are shaking. I mean really shaking, and the teeth are rattling in his head like a machine-gun. Granny.

NO RESPONSE, LOOKS TO GRANDFATHER.

Granddad. What are all them things in your face? Like flints. Making you look like a maze puzzle in a comic. Can you help the lost little girl make her way from the hollow eye back to her Granny's in the forest? There are three hundred blackheads hidden on the way. A glass eye bonus for each one you spot. Flanders? I don't know where that is, Granddad. Is it a laundry, like the White Swan or the Swastika? Do you know where Flanders is, Granny?

SHE TAKES GRANNY'S VOICE.

Flanders is where your grandfather went to in the 1914 war and never came back from. You think that's him in the bed beside us, but that's not your grandfather, that's not Christy Costello, that's only a shadow of a man once was. Oh, that's lovely where you have your feet, child, you are better than a hot water bottle, better than electricity you are. Be a long time waiting for our war hero to warm me up. Stiff with the cold he is. Stiff with the muck in the trenches. And he expects the British to look after him now that they've destroyed him. Dressing him up in a fine uniform only to send him back half a man. But his army pension won't bring back his eye or smooth out his face. Or stop the terrible nightmares that haunt him nightly. Oh, that's perfect where your feet are now. You must have been conceived in front of a rip-roaring fire, child. Now, let's all try and get some sleep. Especially say a prayer that your grandfather sleeps

Peter Sirr

Peter Sirr has published several collections of poems with Gallery Press, the most recent of which are *Bring Everything* (2000) and *The Ledger of Fruitful Exchange* (1995). Until recently he was Director of the Irish Writers' Centre and currently works as a freelance writer, editor and translator in Dublin.

News

A book with the names of the hanged
their names in columns, lightly travelled

certain bodies borne off to the prosecutor's door
or dragged by Volunteers to the waiting professors

the urban renewal of the call
to have the deed brought back from behind the wall

where, for good order, they had placed it
And it was granted, again

colourful scenes on the green, in the square
this one strangled and thrown upon the pyre

this one and his brother strung up yesterday
a harvest, recently, of botched bones

where the inexpertly hacked still lay
deep in the city   and today

alerted by radio to the buzzing of cranes
slowly winching the five victims up

'God is great!', the cranes mounted
on the backs of trucks, someone's

glad thought   always this   always those at the back
with toddlers, straining for a better look

from bathroom to breakfast
a thin span of rage, then ...

a building crew waits
for the return of the cranes

a rope holds, a crowd
disperses, gathers again

William Wall

William Wall is the author of three novels, *Alice Falling* (2000), *Minding Children* (2001) and, most recently, *The Map of Tenderness* (2002), as well as a volume of poetry, *Mathematics & Other Poems* (1997) and many short stories. He was born in Cork in 1955.

In this excerpt Edward Salter lands with his battalion at Sedd-el-Bahr on the Gallipoli peninsula in April 1915. The Gallipoli Campaign was the brainchild of Winston Churchill – its prosecution, despite all military advice to the contrary, was largely due to Churchill's arrogant self-belief. Salter is a classical scholar and the quoted text is from his fictional memoir 'Hoplite & Hotchkiss Gun'.

The Singular Thing
from an unfinished novel

They were marched on board the ancient steam collier *River Clyde*, proceeding to a point just off the island of Tenedos where the attack convoy was to assemble. It did not escape Edward's notice that this was the island the Greeks withdrew to when they left the horse on the Trojan shore, and from here they saw the treacherous Sinon's beacon telling them the gates were open and the defence undone.

The *River Clyde* was to be the new Trojan Horse and on the morning of the attack it was run aground on a stony beach near the desolate Cape Helles.

Of the first two hundred men onto the gangway that fatal morning, one hundred and forty-nine would be killed and thirty wounded. Perhaps, he sometimes remarked with characteristic irony, as they were bleeding or drowning the ringing words of their brigade commander would come to them as consolation: 'Fusiliers, our brigade has the honour of being the first to land.'

An uncompleted poem – considered, to judge by his markings, but rejected for his memoir *Hoplite & Hotchkiss Gun* gives the colour of his feelings:

As we were dying, bleeding, drowning
Nestor weeping, Ajax frowning,

Torn by bullets was the sky,
Torn the sea by Turkish shell:
Troy was beating out its panoply
Across the Dardanelles.

'I saw a man's brains shot away in front of me,' he wrote, 'as he made for the lighters which the navy boys had anchored between us and the shore to make a path for the attack. Shortly after I tasted blood in my mouth and found my eyes clouded. On wiping my sleeve across them I was surprised to find blood and grey matter on it. I concluded that my face had been splashed because I could feel no pain and there was no diminution of my powers. I afterwards realised that it came from the man I had seen die, but at the time my thoughts had been of the impossibility of reaching the shore whilst the Turks concentrated heavy fire on the narrow passage across the lighters.

'I threw myself into the water and promptly began to drown, dragged down by the sixty pounds weight of my pack. By a stroke of luck I managed to find my bayonet and cut the straps before sinking for the last time, and afterwards to swim ashore and take cover behind a low tufted mound with three or four others. I had lost my rifle but had no difficulty finding another as the beach was already littered with corpses. The noise was horrific, and looking back I could see that the water was a perfect mess of exploding shells and bodies. Ahead, the terrain looked impassable, and it was plain that the Turks were dug in behind sound defensive positions and their shooting – much derided on board ship on our way here – looked remarkably accurate from the uncertain cover of a mound of sand and grass.

'By evening the water was red with blood to a distance of fifty yards off and I thought of Homer's fabled "wine-dark sea" and wondered if this was it after all. The beach behind was littered with casualties, most of them dead: and those that were not dead had no chance of crawling to safety for the Turk shot anything that moved.

'Sometime that night, moving along the hastily dug and dangerously shallow sap by the light of the star-shells, I stumbled on the body of a man who had been shot through the eye. He was a Turk, with a heavy face and huge peasant's hands, and he stared up at me with contempt out of his one baleful eye. I thought of the humiliated Cyclops, blinded by trickery: surely you sorrow for the eye which an evil man put out, cursed be he and all his fellows. From that moment onwards I knew our enterprise could not ultimately succeed. We had been anathematised and Poseidon the earth-shaker would keep us from our country for many's the weary year.

'Our orders were simple. We were to charge the Turkish defences with fixed bayonet: to drive them from the height and possess it; to hold it at all costs. The orders were devised without any intelligence of the terrain, the extent of the barbed wire entanglements we could expect to encounter, the strength of the enemy's position, their state of training, or the condition of what was left of our own battalion. It may seem like hindsight to remark that it was ill-conceived and unsoldierly, but Major-General Sir Aylmer Hunter-Weston could hardly have appreciated the fragile hold we had of the beach from the comfort of his cruiser out in the wine-dark straits.

'Noise is the singular thing: not the dangers posed by shot and shell, nor the imminence of one's own demise or that of one's fellows. The noise fills the head and is reinforced by the physical disturbance of the air, earth and the sea. It is true, that after hours days weeks of bombardment, attack and counter-attack, the whole thing becomes like background music. Still, no soldier ever loses his consciousness of the *pattern* of firing. A minor change in that brings him to attention very smartly and questioning glances pass along the sap: Where is the firing? Who's up now? What is the Turk doing?

'Here are the sounds that live in memory: rifle-fire like a hundred cricketers at practice, volleying and cracking; the peck-peck of a Mauser; the scream of naval shells coming in; the rustling of near-spent bullets; the sound of sentries in pairs, yarning; the banging of mountain guns and field pieces during an attack, like living inside a tin drum beaten at both ends by crazed giants; the curious whizz on a descending scale of shrapnel shells. And once, when we had exploded a mine under a salient in the Turkish line, and we had charged in to take possession of it before they could, the firing ceased for a few seconds on both sides and close by my head, where timber was mixed with clay and sand, I heard the light, lively ticking of a watch.

'At one point we had taken a Turkish trench. The fighting had been bloody, our charge somehow coinciding with some changes in their dispositions, and we were upon them with bayonets before they could respond. They had, quite properly, turned tail and abandoned the position to us, but we knew that they would be regrouping not far away. They were very valiant fighters, those Turks.

'It is notoriously difficult to defend the back of a trench – the *parados* – and we were working feverishly to shift sandbags to that side and hollow out a fire-step. At the same time our stretcher-bearers were coming up to get the wounded out. It was during these moments of frenzied activity that I noticed the trench had been cut through layers of tiles and pottery at a depth of about five feet from the surface. In fact, there were faded tesseræ underfoot and our shovels were pulling pieces out as we worked. I could not tell how old they might be, whether from the recent past or some ancient settlement, but the latter seemed not at all unlikely for the entire sweep of the Troad and the noble mound of the ancient sacked city itself were in view behind us across the water. I had time, I know, to point them out to a friend who shared my interest.

'We agreed that if the position held against the counter-attack we would spend a little time examining them and perhaps remove some pieces as souvenirs. At that point the enemy machine-guns fell silent and the cry of "Allah!" was raised. The Turks were upon us and very soon we were driven out of that trench never to return. The entire position was undermined sometime during the summer and the trench and its tesseræ blown sky-high.'

David Wheatley

David Wheatley was born in Dublin in 1970. He has published two books of poetry with Gallery Press, *Thirst* (1997) and *Misery Hill* (2000).

The Gasmask

If I still had the use of my mind I'd call it insane.
When I got in the taxi he took me, as we drove home,
on a detour through a pet theory of his
that went something like: *cells of them, one big plot,*
already over here, anthrax, Saddam, our boys,
only language he understands, gas attack,
one in the attic, my granddad's, Geiger counter ...

after which the next thing I heard was a grunt, or
was it me slithering down the seat in the back,
then a honk rocking the cab with the noise
as he turned and a pendulous, rubber snout
loomed at me from his wall-eyed elephant's face:
he'd already put it on and I had to tell him,
'Sorry, I can't understand a word you're saying.'

Enda Wyley

Enda Wyley was born in Dublin in 1966. She has published two
collections of poetry, *Eating Baby Jesus* (1994) and *Socrates in the
Garden* (1998). A third collection of her poetry *Diary of a Fat Man* is
forthcoming.

After Reading a Poem by Yehuda Amichai

I did not know you, little Ruth,
and yet, here in this old house,
in the village of Castlegregory –
far from my Dublin home –
I open a book on my rickety desk
and find your life, written by the poet there.

It is mid-February. Crows peck for slugs
in the hardened soil, the day is frail
as sycamores against the cold sky,
the cold sea beyond – under my window
I see a white cat flutter free like snow,
from the farmer's hungry wire fence.

And on the page, the memory of you
burnt to death in the camps. You were just twenty –
if alive, you'd be old now, a woman nearing eighty –
but once little Ruth, a child not knowing just how
they would take you away, stamping your bright soul
with stars, cruel insignias of their own hatred of you

and those you loved, those that belonged to you.
Where is your life that might have been?
The poet goes out with his children now,
the children you never met,
gathers mushrooms in the very forest
he remembers planting with you as a child.

I did not know you, little Ruth, but your life
is still on this earth, a cluster of precious words
shot like last night's meteorite from the burning
memory of your friend into my waiting mind.
We are bundled together, the poem about you
is also about me – this is how love lives on

quietly, unsuspecting that the turn of a page
would reveal so much, even here in a Kerry house,
at the turn of a new century, far from where
you lived, far from where they killed you,
far from where all memory starts.
The season changes. The trees are on fire

with summer warmth, the cat snoozes
against the heat of the fence's wooden post
and the azure waves are calling us.

We are leaving the dark forest, little Ruth,
wild flowers a tangled hope in our hair,
and on the beach, my footsteps follow yours
happily, way out to the beginning of the skies.

13 February 2003

Ann Zell

Ann Zell was born and raised in the USA. She has lived in Belfast since 1980. Her first collection, *Weathering*, was published by Salmon Press in 1998. Her second collection is due for publication in 2003.

Behind Glass

The direction of the wind
can be known from the lift of the sand
the progression of clouds
the curve of the grass
a silver skim
crossing the skin of the water.

The causes of the weather
are explained each night
in animated maps.

From where I sit, behind glass,
animated maps will show
the weather of the war –
warm fronts, cold fronts,

clusters of precipitation
falling on the citizens of Iraq.

The causes of the war
will be explained in lies.

A Note on The Irish Anti-War Movement

The Irish Anti-War Movement is an alliance of anti-war groups, trade unions, political parties, community groups and thousands of individuals who organised across Ireland to oppose the current US-led war.

War on Iraq will needlessly claim the lives of innocents and increase instability and conflict in the world. It will create the conditions for future wars.

Over one hundred million people died in war in the twentieth century. With modern military technology, war in the twenty-first century holds the potential for human destruction and environmental destruction on an even greater and more frightening scale.

The vast majority of people in Ireland and across the world are appalled at the current war. On 15 February 2003 tens of millions across the globe took to the streets to say no to war. This global protest was proof that there is an alternative to war based on the solidarity of peoples across the world.

The Irish Anti-War Movement was at the centre of organising that protest in Ireland. We are working to establish an organised anti-war movement in every corner of Ireland. We want to play our part in building the international movement to stop all wars. We believe that through the actions of millions of ordinary people co-ordinated together it is possible to stop war. We would like to thank the contributors, editors and publishers of this anthology for their support in building that movement. We hope that this book will encourage you, the reader, to join us and to one day make war a thing of the past.

Richard Boyd Barrett, Chair

...ke to join the Irish Anti-War Movement please send a ...al order made out to 'Irish Anti-War Movement' along ...r name, address, phone and email to:

Irish Anti-War Movement
61 Stewart Hall
Parnell Street
Dublin 1

Membership is €5 if you're unwaged, €10 if you're waged, or as much as you can afford.

The anti-war campaign involves large expenses, particularly for printing costs and room hire. Contributions are very welcome. They can be sent to the above address or made directly to:

The Irish Anti-War Movement
Bank of Ireland
34 College Green
Dublin 2
a/c number : 39640902
sort code: 90-07-89